Philip K. Groom, Byron B. Lamont
Plant Life of Southwestern Australia. Adaptations for Survival

Philip K. Groom, Byron B. Lamont

Plant Life of Southwestern Australia

Adaptations for Survival

Managing Editor: Katarzyna Michalczyk

DE GRUYTER
OPEN

Published by De Gruyter Open Ltd, Warsaw/Berlin
Part of Walter de Gruyter GmbH, Berlin/Munich/Boston

ISBN: 978-3-11-037016-4
e-ISBN: 978-3-11-037019-5

Bibliographic information published by the Deutsche Nationalbibliothek
The Deutsche Nationalbibliothek lists this publication in the Deutsche Nationalbibliografie; detailed bibliographic data are available in the Internet at http://dnb.dnb.de.

Managing Editor: Katarzyna Michalczyk

www.degruyteropen.com

Cover illustration: © Peter Mioduszewski

"Owing to its arresting beauty, the vegetation of Western Australia arouses both interest and affection"

Ludwig Diels (1906)

Acknowledgments

Much is known about the southwestern Australian flora. The impetus for preparing this book was the need to compile this vast knowledge base into one publication. We would like to thank members of the Department of Environment and Agriculture (and its predecessor the Department of Environmental Biology), Curtin University, Western Australia, for their support and encouragement during the writing phase, especially Professors Jonathan Majer and Jacob John. Special mention goes to the thousands of students we have taught in the undergraduate unit 'Flowering Plants/Australian Flora' (and their predecessors) for inspiring us to make this book a reality.

Most photographs and drawings were sourced from the authors' personal collection. Allen Lowrie graciously provided photographs of the wasp-pollinated orchids, in addition to images of other orchids and *Drosera* species. Mark Brundrett kindly provided additional orchid photos including the monotypic genera and the underground orchid. Special thanks to Jiri Lochman of Lochman Transparencies for providing several photos from his vast photographic collection. William Archer allowed us the use of photographs from his online collection *Esperance Wildflowers*. We also thank Haylee D'Agui, Geoff Burrows, Melissa Darling, Katherine Downes, Aaron Gove, Geoffrey Groom, Peter Mioduszewski, William Parkinson, Allan Tinker and Stephen van Leeuwen for providing additional photographs. Katherine Downes is thanked for providing information on fire ephemerals and soil-stored seeds. Hervé Sauquet provided the Proteaceae chronogram that was then reanalysed for fire-proneness and fire-related traits and redrawn by Tianhua He. Maps and tables were compiled from published journal articles or books, the sources of which are acknowledged in their associated captions. Wesley Lamont is thanked for redesigning the maps and all line drawings. Christine Cooper and Tim Lardner assisted with access to relevant references. Alex George, Neville Marchant, Laco Mucina and Ross Young graciously commented on earlier drafts. Brian Heterick assisted with ant identifications. Special thanks to Fran, Byron and Zoe from Dome Maylands for all their encouragement and support during the later drafts of the book.

We dedicate this book to Peter Mioduszewski, a colleague and friend who had a passion for the southwestern Australian flora and natural history in general. He was well respected by his peers, and the students he assisted as a botanical technical officer at Curtin University. Peter passed away suddenly in late 2012. In his memory we selected one of his photographs for the cover page.

We began writing this book in 2010, the International Year of Biodiversity.

Contents

Prologue

"Knowledge of the plant life of Western Australia was very limited at the time of my visit. While the floral elements were well known, no studies had been made on their inter-relationships in the field....little was known about the conditions that determined the inner relationships of the endemic flora. These aspects were considered as being fundamentally important...

...Also the country's unrivalled richness in species could be expected to prove rewarding to those who studied its vegetation".

Ludwig Diels (1906)

Extract from a translation of the book "Die Planzenwelt von West-Australien südlich des Wendekreises"

In 1906, Dr Friedrich Ludwig Diels (1874–1945), Director of the Munich Botanic Gardens, published his book *Die Planzenwelt von West-Australien südlich des Wendekreises*, translated meaning *Plant Life of Western Australia South of the Tropics*, based on his 1900-1902 travels with fellow botanist, Dr Ernst Pritzel. The uniqueness of the Western Australia flora was already recognized prior to this book's publication but Diels' work is considered seminal to our current understanding of the distribution and general composition of the flora; however, his work has not been easily accessible, as a translated version only became available in 2007.

P.1 Southwestern Australia and its Functional Biodiversity

Southwestern Australia (hereafter referred to as the 'SouthWest') – defined as the area south of a line from Shark Bay in the north to Eyre in the south-east – is unique among Australia's bioregions because of its high species richness and levels of endemism and exceptional functional diversity. The SouthWest is Australia's only recognized biodiversity hotspot among 25 regions world wide (Myers *et al.*, 2000). The SouthWest is also special in that it is one of only five regions that experiences a mediterraean-type climate characterised by cool, wet winters and summers that are hot, dry and prolonged.

Biodiversity can be defined as the level of variability present among organisms and their life processes within a reference area, and includes composition, structure and function at the scales of genes, species, communities and ecosystems (Noss, 1990; Lamont, 1995). The functional component of biodiversity in the SouthWest flora relates directly to trait diversity resulting from macro- and micro-evolutionary trade-offs between form (e.g. morphology, structure) and function (e.g. physiology, protection) shaped by climatic, edaphic, pyric, hydrological, geological and biotic pressures that began over 100 million years ago and intensified over time.

Thus, it is not just the number of plant taxa, diversity of habitat types and ancient, weathered landscapes that sets the SouthWest apart from other Australian bioregions. The flora also exhibits a diverse range of morphological and physiological adaptations that aid its survival in extremely nutrient-impoverished, drought- and fire-prone environments and subject to intense herbivory and granivory by such unique animals as strong-billed black cockatoos. It also lacks the long-tongued social bees and humming birds that are such efficient pollinators in other continents but has instead solitary bees and unrelated nectar-feeding birds and small marsupials that can take over the task. Emus and kangaroos are not only voracious plant feeders to which the flora must adapt but these animals can also be utilized for long-distance transport of seeds. The SouthWest flora is characterized by tough, prickly-leaved species that have a persistent seed bank stored either on the plant or in the soil, have modified roots or symbiotic relationships with microorganisms that enhance soil nutrient and water uptake, and a general reliance on fire for long-term survival. These adaptations are not necessarily confined to the SouthWest, they are just exceptionally well developed here.

P.2 Environmental and Biotic Constraints

Our book recognizes that the major environmental constraints to which species in the SouthWest flora have had to adapt can be grouped into three *abiotic factors*: 1) the nutrient-impoverished soils, 2) seasonal drought and heat, and 3) recurrent crown fires, and two *biotic factors*: 1) absence of highly efficient pollinators and dispersal agents, but 2) presence of highly efficient herbivores and granivores (Fig. P.1). Our emphasis is on environmental factors that constrain growth and reproduction and to which species have had to adapt through the processes of genotypic variation and directional selection if they are going to survive, followed by stabilizing selection among the adapted core of survivors. Few taxa were able to adapt to such severe constraints but those that did underwent exceptional rates of speciation with remarkable proliferation of their adapted traits leading to a limited number of clades at the levels of widespread genera and families but remarkably high numbers of endemic taxa at the species level (Hopper & Gioia, 2004).

Thus, the family Proteaceae, with a 115-million-year history, is distributed throughout Australia, Africa, South America, New Zealand, Asia, New Guinea and New Caledonia but 800 of the 1,400 species in the family occur in the SouthWest with over 95% confined here (Cowling & Lamont, 1998). It is no accident that >90% of these species have novel dispersal agents (ants, emus, willy-willies, Chapter 10), fire-responsive seed storage and release (with a 90-million-year history, Chapters 1, 10, 11), life cycles adapted to recurrent crown fire (Chapter 2), specialised root clusters for enhancing nutrient uptake (with a 90-million-year history, Chapter 6), a wide range of pollinators, including the novel use of bees, marsupials and birds (Chapter 8), and unusual breeding systems including protandry and flower colour change (Chapter 7), among the world's most scleromorphic, long-lived and spinescent leaves as a response to impoverished soils (with a 100-million-year history), and later exposure to recurrent drought and strong light (with a 15-million-year history, Chapters 3, 9), and novel granivores such as moths and cockatoos that have involved innovative solutions such as crypsis and woody fruits (with a 20-million-year history, Chapter 9). Paradoxically, the seeds of SouthWest Proteaceae store among the highest levels of N and P in their seeds of any worldwide but are accumulated from foliage and soils whose levels of these nutrients are among the lowest in the world (Hocking, 1981; Kuo *et al.*, 1982; Hocking, 1986; Pate *et al.*, 1986; Groom & Lamont, 2010).

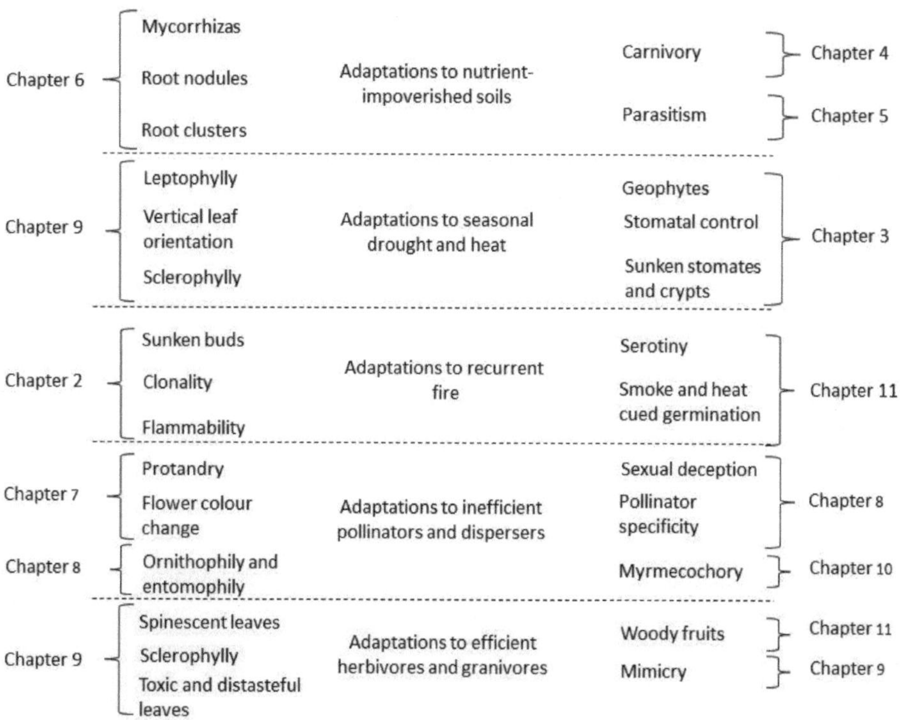

Fig. P.1: Adaptive responses to the major constraints on survival, growth and reproduction of SouthWest plant taxa, as they relate to chapters within this book.

Both the large seeds and scleromorphic leaves share the constraint sequence through time of nutrient impoverishment beginning in the high rainfall climates of the

Upper Cretaceous followed by increasing drought and seasonality in the Miocene with fire a pervasive but increasingly restrictive selective agent throughout. While climates in the Paleogene are considered to have been even wetter than the Cretaceous, it is clear that the climate was seasonal and still fire-prone at this time (Lamont & He, 2012). These same climatic effects oscillated at about 100,000-year intervals throughout the 2.6-million-long Quaternary (Rognon & Williams, 1977; Byrne, 2008) and hastened the extinction-adaptation cycle to produce the unique SouthWest flora we see today. Thus, while we accept the importance of poor soils (Hopper, 2009) and fire (Mucina & Wardell-Johnson, 2011) of previous models in shaping this flora, we cannot accept that climatic stability has had a key role as proposed by Hopper (2009). Increasing and fluctuating seasonality and aridity over time have had a major impact on plant traits (Lamont *et al.*, 2002) but also on the other two constraints, especially the formation and distribution of the nutrient-impoverished laterite and sands and exposure of the parent granite so characteristic of the SouthWest (Glassford & Semeniuk, 1995), and the intensity and frequency of fire (Pausas & Keeley, 2009).

P.3 Our Book

The Western Australian Herbarium has adopted a systematic approach to the classification of vascular plants, based on current research conducted by the Angiosperm Phylogeny Group, a global partnership among plant scientists interested in understanding the relationships between plant families and orders. This revision, known as APG III (The Angiosperm Phylogeny Group, 2009), involves some family-level changes. To maintain consistency with the Herbarium's move towards the APG III's taxonomic classification, this book also follows this system. We have also endeavored to provide current species and plant family names according to *Florabase*, the Western Australian Herbarium's authoritative online species database. Several myrtaceous genera have now been merged into *Melaleuca* (Craven *et al.*, 2014), including *Beaufortia, Calothamnus* and *Eremaea*, all endemic to the SouthWest. We have elected to maintain these genera because of ecological differences between and within these genera and *Melaleuca sensu stricto*, and also readers may be unfamiliar with the new nomenclature resulting from these recent combinations.

The title of our book—*Plant Life of Southwestern Australia*—is in part a tribute to Diels' monograph. Whereas Diels' book focused on describing Western Australia's temperate and semi-arid flora according to its composition, distribution, and structure, our book is about the flora's adaptations, hence the subtitle *Adaptations for Survival*. Our book begins with a general introduction to the evolution and diversity of the flora, then details adaptations that are in response to nutrient-impoverished soils, recurrent fires and summer drought. Finally, strategies or morphologies designed to maintain or ensure species survival that relate to pollination, plant leaves and seeds are discussed. While compiling this book, we re-examined the available literature to provide a comprehensive up-to-date overview of the topics discussed.

Just like Ludwig Diels' book, we hope that our book will "*prove rewarding*" to those who have an interest in southwestern Australia's unique flora, as well as providing insights into the adaptations that the native flora has developed to survive what can be a persistently harsh environment.

1 Evolution and Diversity of the Flora

Land plants arose almost 500 million years ago (Ma). Those with a vascular system (lycopods) appeared at least 450 Ma, followed by the ferns with true leaves rather than scales at least 400 Ma. Seed-bearing plants originated about 430 Ma giving rise to the gymnosperms about 330 Ma, including the cone-bearing cycads in one clade (phylogenetic group) and the conifers in another.

The latest research based on rates of change in DNA structure (the molecular clock) puts flowering plants arising at least 134 Ma but it may have been up to 248 Ma depending on the true affinity of different fossils and the accepted rates of molecular change and whether a group is only recognised once it starts to separate into its daughter groups (called the crown) or at its base where it remains undivided or only a single-ancestor can still be identified at the molecular level (Clark *et al.*, 2011). This puts angiosperm origins in the Permian Period, or as early as the Ordivician, much earlier than once thought. Nevertheless, once the angiosperms appeared they gradually outcompeted all previously existing groups of plants, certainly by the end of the Cretaceous, 65.5 Ma. This was apparently due to their rapid growth rates and superior colonizing ability. Speciation of the SouthWest angiosperm flora was particularly rapid during the Miocene to Pleistocene, a time of great climatic upheavals with a general drying and cooling trend from 15 Ma (Fig. 1.1), including a period of severe aridification 2–5 Ma (Dodson & Macphail, 2004). Many plant lineages that were once widespread within temperate Australia, now have separate taxa in the SouthWest and southeast/eastern Australia because of the biogeographical barrier imposed by aridification and the elevation of limestone sediments in the current Nullarbor Plain, beginning 14 Ma (Crisp & Cook, 2007).

1.1 A Radiating and Hyperdiverse Flora

Elements of the Australian flora were already present in the Upper Cretaceous, 90 Ma, mostly confined to rainforest dominated by conifers but also in fire-prone, sclerophyll pockets (see Fig. 1.5 – Proteaceae phylogeny). The large asteroid that struck the Earth in the Gulf of Mexico 65.5 Ma, causing an abrupt end to the Cretaceous Period, was an important event for the world's biota, including the SouthWest. Within 4 million years (61–62 Ma), key taxa in the SouthWest flora such as *Banksia* (Proteaceae), *Eucalytpus* (including *Corymbia*) (Myrtaceae) and the Australian Restionaceae emerged (Fig. 1.1). All these groups are adapted to nutrient-impoverished soils and highly fire-tolerant. Eucalypts are at present the dominant overstorey species in temperate Australian woodlands and forests and they further enhanced the flammability of the flora through the aromatic oils in their leaves and stems though it is thought that initially they evolved to deter herbivores (Crisp *et al.*, 2011). Eucalypts also have a distinctive epicormic resprouting structure (Chapter 2; Burrows, 2013) that remained unchanged throughout the subsequent Cenozoic Era to the present time (Crisp *et al.*, 2011).

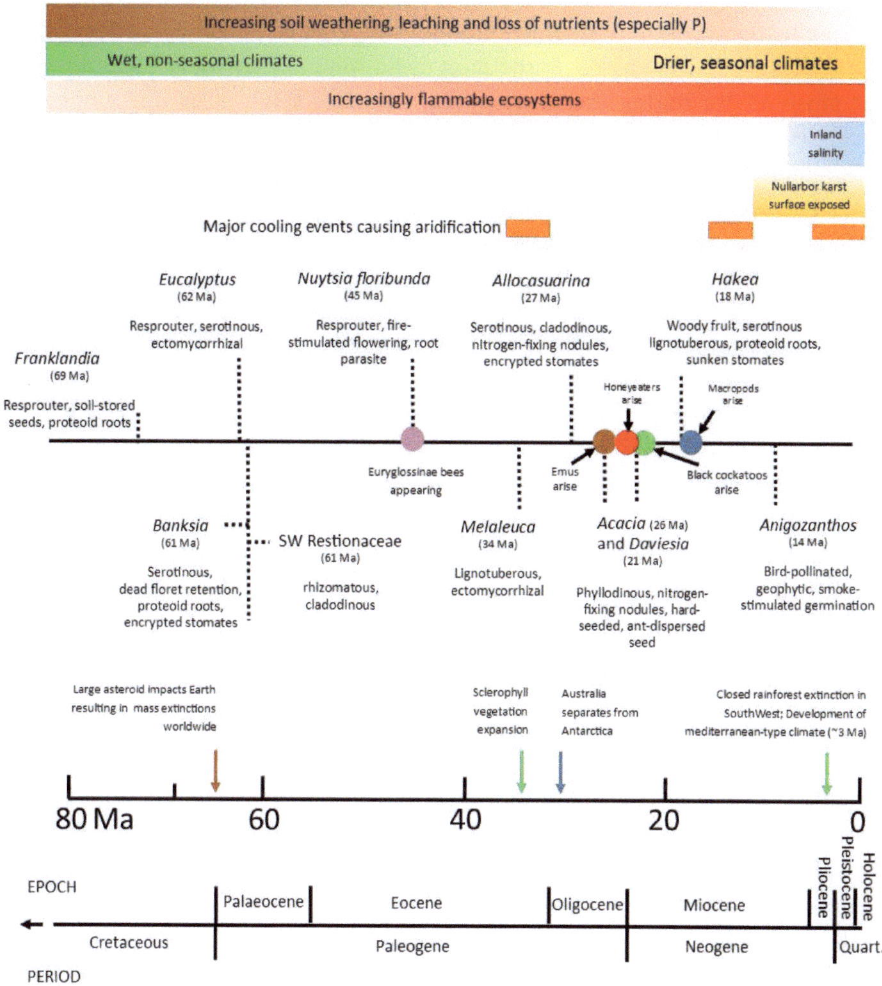

Fig. 1.1: Appearance of extant species-rich and iconic SouthWest taxa in Australia since the late Cretaceous Period compared with the occurrence of biotic and environmental agents of selection. Euryglossin bees and honeyeaters are pollinators of the flora (Chapter 8) whereas black cockatoos have influenced the evolution of woody-fruited species (e.g. *Hakea*) (Chapter 11). Geological time compiled from The Geological Society of America Geological Time Scale (2012, v. 4), and key events obtained from Byrne *et al.* (2014). Formation of Nullarbor Plain and aridification events since 35 Ma obtained from Crisp & Cook (2007). Information obtained from time-calibrated molecular phylogenies described in Crisp *et al.* (2004) [*Allocasuarina*]; Vidal-Russell & Nickrent (2008a) [*Nuytsia*]; Hopper *et al.* (2009) [*Anigozanthos*]; Prideaux & Warburton (2010) [Macropods]; He *et al.* (2011) [*Banksia*]; Crisp *et al.* (2011) [*Eucalyptus, Melaleuca*]; White *et al.* (2011) [Calyptorhynchinae cockatoos]; Miller *et al.* (2013) [*Acacia*]; Almeida *et al.* (2012) [Euryglossinae bees]; Litsios *et al.* (2013) [Restionaceae]; Baker *et al.* (2014) [emus]; Joseph *et al.* (2014) [honeyeaters]; *Franklandia* and *Hakea* from Fig. 1.5.

Tree root-parasites are basal within the hemiparasitic Santalales order and the best known in the SouthWest is the monotypic *Nuystia floribunda* (Loranthaceae) that arose 45 Ma during the Eocene (Fig. 1.1). This relict is unique among the Santalales in having wind-dispersed fruits (all others have succulent fruits that are a feature of rainforest species), precocious fire-stimulated flowering from fire-blackened stems, tough, semi-terete leaves, and rhizomes reaching 100 m in length. These features suggest that at least parts of the SouthWest during the Eocene had a highly seasonal climate and were subject to frequent fire (Lamont & He, 2012). Parasitism in the Loranthaceae evolved from terrestrial to aerial 28 Ma, with rapid diversification occurring during the Oligocene, an epoch of global cooling and temperate woodlands and grasslands (Vidal-Russell & Nickrent, 2008a). Today Loranthaceae is dominated by aerial hemiparasites, though only *Ameyma* and *Lysiana* are present (and widespread) in the SouthWest (Chapter 5). Both genera are pollinated by honeyeaters (Meliphagidae) that originated 23.5 Ma and radiated strongly between 15 to 5 Ma (Joseph *et al.*, 2014; Fig. 1.1). The frugivorous mistletoe bird (*Dicaeum hirundinaceum,* Dicaeidae) is a major seed disperser (Yan, 1993a), although this species did not colonise Australia until the Pleistocene, possibly as late as the Holocene (Reid, 1987), arriving from New Guinea and the surrounding islands (Nyári *et al.*, 2009) and was already adapted to its mistletoe berry diet. This bird never reached Tasmania.

The Ecocene also saw the origin of solitary stingless bees in the Australian endemic subfamily Euryglossinae (Colletidae) (Fig. 1.1) and the Australian biodiverse subfamily Hylaeinae (48-54 Ma) (Almeida *et al.*, 2012) that are important pollinators of small-flowered species in Australia. Emus (large flightless bird) (*Dromaius novae-hollandiae*: Dromaiidae) appeared 27 Ma (Fig. 1.1) and became important long-distance seed dispersers.

Major taxonomic groups were well established in the Australian flora by the time Antarctica split from Australia as it pulled north, beginning in the West in the early Eocene and finishing with Tasmania during the early Oligocene, leaving the flora to radiate in isolation from other continents, as the climate became colder and drier with Australia's drift into the mid-latitudes. The Casuarinaceae (*Allocasuarina, Casuarina* and *Gymnostoma*) radiated about 47 Ma (Crisp, Cook & Steane, 2004), with the *Allocasuarina*/*Casuarina* clade developing about 39 Ma, with their leaves reduced to minute scales, encrypted stomates hidden within cladodinous grooves, and serotinous woody cones (Crisp & Cook, 2013). Speciation in *Allocasuarina* appears to have accelerated about 25 Ma dominating the expanding temperate sclerophyllous floras of the SouthWest and southeastern Australia (Crisp *et al.*, 2004).

The greatest rates of speciation occurred during the long Miocene Epoch (Fig. 1.1), as the climate became drier and more seasonal. Rates of trait proliferation were highest in both directions at that time. For example, the rates of evolution of species with serotiny and non-serotiny, retention and shedding of dead florets and leaves, and clonality and non-clonality among the genus *Banksia* were greatest then

(He *et al.*, 2011). This indicates that a great range of habitats and climate extremes was present at that time and that the Miocene was an important period in the evolution of the SouthWest flora (and Australia generally). The only exception among banksias was the thickening of the mantle of dead florets that finally covered the fruits that peaked in the Pliocene (2–5 Ma), either as a response to increasing reliance on fire to melt the resin sealing the fruits or to improve crypsis from the increasing presence of granivorous black cockatoos (Chapter 10). The dominance of the Proteaceae within the SouthWest flora throughout much of the Cenozoic (Itzstein-Davey, 2007) is believed to have been a major contributor to the development of the iron/aluminium-rich lateritic soils here (Pate, Verboom & Galloway, 2001) because of the long-term effect of the family's proteoid root clusters that exude organic acids to increase phosphorus uptake efficiency (Chapter 6) with subsequent mobilization and redeposition of iron/aluminium oxides at depth. The concreted laterite was subsequently exposed during the arid periods that followed as the surface sands were blown into the valleys (Glassford & Semeniuk, 1995). During wetter periods even the upland laterite and underlying kaolinitic clay were sometimes washed away to expose the parent gneissic granite, to produce monadnocks that are also a feature of the SouthWest landscape.

The iconic and ecologically-important Australian and SouthWest genus *Acacia* (Fabacae: Mimosoideae) arose during the late Oligocene (Fig. 1.1). Originating 26 Ma (Miller *et al.*, 2013), *Acacia* has radiated to over 1000 extant Australian species, many in the SouthWest. Arid zone species related to *A. victoriae* and *A. pyrifolia* represent an early branch in the phylogeny of Australian acacias, although the estimated age of these taxa is less than 7 Ma (Miller *et al.*, 2013). *Acacia* species possesses either bipinnate leaves or phyllodes (flattened petioles that function as leaves) in mature plants. All seedlings possess compound leaves which suggests that this is the ancestral leaf type. Bipinnate leaves occur in the sections *Botrycephalae* and *Pulchellae* and are rare compared with phyllodinous species that may dominate the driest parts of the SouthWest. The *Pulchellae* radiated 17–18 Ma, whereas the complex *Botrycephalae*/*Phyllodineae* clade (where compound and phyllodinous species are interspersed) originated between 7.5 and 13.5 Ma and occurs in wetter areas (Miller *et al.*, 2013). Phyllodes may be simple or fasciculate (clustered), although fasciculate species are not common (5% of the genus) and not confined to any Section (Maslin, 2014). Another fabaceous genus, *Daviesia*, radiated 21 Ma, and consists of phyllodinous and cladodinous species and some species developed roots with anomalous secondary thickening (cord roots) (Pate *et al.*, 1989; Crisp & Cook, 2003).

Modern emus (*Dromaius novae-hollandiae;* Dromaiidae) appeared at 26 Ma though their ancestors (even larger, flightless birds) can be traced to 78 Ma (Baker *et al.*, 2014). These voracious omnivores pluck foliage, flowers and fruits (and insects) from almost any plant available (Calviño-Cancela *et al.*, 2006) so they have probably had a role in the evolution of anti-herbivory traits, such as spiny leaves and mimicry. Interestingly, SouthWest emus have had sufficient time to develop immunity to the toxic fluoroacetates produced by many *Gastrolobium* species here

(Chapter 8). Macropods evolved much later at 17 Ma (Prideaux & Warburton, 2010) as the Miocene became drier and more open vegetation, especially grasslands, developed. Kangaroos and wallabies browse on a wide range of species and young plants in particular are highly vulnerable, especially if they are of above average height and have a grasslike appearance even when they are actually dicotyledons (Rafferty *et al.*, 2005; Parsons *et al.*, 2006; Rafferty & Lamont, 2007). Chemical defences, such as phenolics (tannins), are effective at the seedling stage (Rafferty *et al.*, 2005, 2010; Parsons *et al.*, 2005; Parsons *et al.*, 2006). Spinescence is only strongly developed in mature plants when it can be quite effective against kangaroos, preventing new axillary growth from being reached (Hanley *et al.*, 2007). Thus, macropods may have been a selective force in the evolution of these traits over the last 15 million years. Interestingly, rat kangaroos (bettongs in the SouthWest) evolved much earlier at 28 Ma (Prideaux & Warburton, 2010). As essential dispersal agents of ectomycorrhizal spores of 'truffle' fungi (Lamont *et al.*, 1985) they may have played a key role in the evolution of both plants (e.g. eucalypts, peas) and higher fungi here.

Hakea (Proteaceae) originated 18 Ma (Fig. 1.1; 1.5) and is a genus of remarkable functional diversity, storing seeds in serotinous woody fruits held in a crown of broad (with or without leaf spines), needle-like (with apical spines) or strap-like leaves. We propose that black cockatoos have had an important role in leaf and fruit trait development to deter avian granivores that is explored in Chapters 9 and 11. Preliminary work on a time-calibrated *Hakea* phylogeny (T. He, unpublished) shows that needle-leaved species arose from broad-leaved ancestors 14 Ma (consistent with decreasing rainfall and increasing pressure from large herbivores and granivores), and the extraordinarily large fruits of the *H. platysperma* and *H. pandanicarpa* clades evolved 10 Ma (consistent with increasing black cockatoo activity). Bird (honeyeater) pollination arose 14 Ma from insect-pollinated ancestors (Hanley *et al.*, 2009) at a time when these birds were beginning to radiate. Fruit mimicry appeared in the *H. prostrata*/*ruscifolia* and *H. trifurcata*/*lasianthoides* groups at about 7 Ma. *H. clavata*, with the thickest leaves of all sclerophyllous species anywhere in the world (4 mm) and confined to rock crevices on monadnocks in the Esperance area, arose 6 Ma. Resprouting and serotiny are ancestral in *Hakea* evolving from a non-sprouting, non-serotinous lineage under rainforest conditions. Many of the most speciose extant SouthWest genera that diversified during the Miocene contain both fire-killed (but fire-stimulated germination) and resprouting species, supporting the importance of fire as an evolutionary force in the adaptive radiation of the SouthWest's hyperdiverse flora.

Arising in the mid-Miocene, *Anigozanthos* (Haemodoraceae) is the most striking herbaceous genus in the SouthWest (*A. manglesii* is also Western Australia's floral emblem). Not only is it bird-pollinated (Hopper & Burbidge, 1978), taking advantage of the honeyeaters now firmly established here, but its germination is stimulated by smoke, the active ingredient of which is glyceronitrile (Downes *et al.*, 2014).

Apart from the evergreen *A. flavidus* that occurs in wetlands, it usually dies back over summer-autumn. Most species are short-lived taking full advantage of the extra nutrients released after fire, and *A. humilis* will flower the first spring after fire.

1.2 Evolutionary Constraints and Responses

While currently poor soils, fire proneness, and seasonal aridity and high temperatures coincide in the SouthWest, it is difficult to separate primary from secondary constraints. However, we support the notion that nutrient impoverishment, especially of P, was the instigator of change and that this had a subsequent cascading effect on other constraints. Greatest leaching and loss of P would have occurred from upland profiles exacerbated by closeness to the bedrock (gneissic granite is a poor source of mineral nutrients), strongly-leached, kaolinitc clay and subsurface laterite with P locked up in insoluble iron and aluminium oxides. Such sites also drain readily and, together with their shallowness, are more drought-prone. Reduced nutrient and water availability lead to smaller plant stature, a trend to evergreenness, and retarded litter decomposition rates. Leaves become increasingly scleromorphic and tannin-filled (primary effect), increasing their longevity and further reducing litter decomposition rates. Without a tree canopy cover, shrubs and small trees are more exposed to full sunlight and higher temperatures. Drought and heat adaptations develop as secondary effects.

Not only are drier, finer foliage and greater litter accumulation more flammable but exposure on uplands makes the vegetation more vulnerable to lightning strikes. Fire adaptations appear as secondary effects. With time, leaching, seasonality and aridity became more generally distributed and the sclerophyllous flora spread over the entire SouthWest. Such traits as carnivory, parastisim, geophyty (retreating to dormant underground structures in summer) and clonality become more common. The temperate rainforest disappeared by the late Pliocene, never to return (unlike eastern Australia).

In a similar way the switch from insect to bird and mammal pollination is a secondary effect that takes advantage of the introduction of more efficient pollinators. Spinescence, crypsis, woody fruits and long-distance dispersal are secondary responses to the later introduction of such formidable omnivores as emus, kangaroos and cockatoos.

Functionally the extant SouthWest flora is extraordinarily diverse with dominant genera developing distinct (and characteristic) traits to survive and compete in what is an overall harsh and climatically-diverse biome (Beard, 1984). Within a given plant attribute, taxa can be placed along a low to high cost gradient or economic biospectrum (Fig. 1.2). The concept that plant traits occur along economic biospectra was proposed by Wright *et al.* (2004) based on the occurrence of worldwide trade-offs in leaf attributes from 'low cost' (thin, soft leaves with a short leaf lifespan) to 'high cost' (thick, hard leaves with a long leaf lifespan) (Fig. 1.2), also known as 'fast-slow' trade-offs at the leaf, stem and root levels (Reich, 2014). Part of the problem with the concept is that costs and benefits cannot be assessed using the same 'currency'. Also costs depend on the currency under consideration. Note for example that scleromorphic leaves are a high cost in terms of carbon but not of N and P and that the benefits in terms of increased fitness far outway the costs of production. These biospectra provide an additional framework that can be used to investigate the significance of broad-scale plant responses to current and past evolutionary pressures.

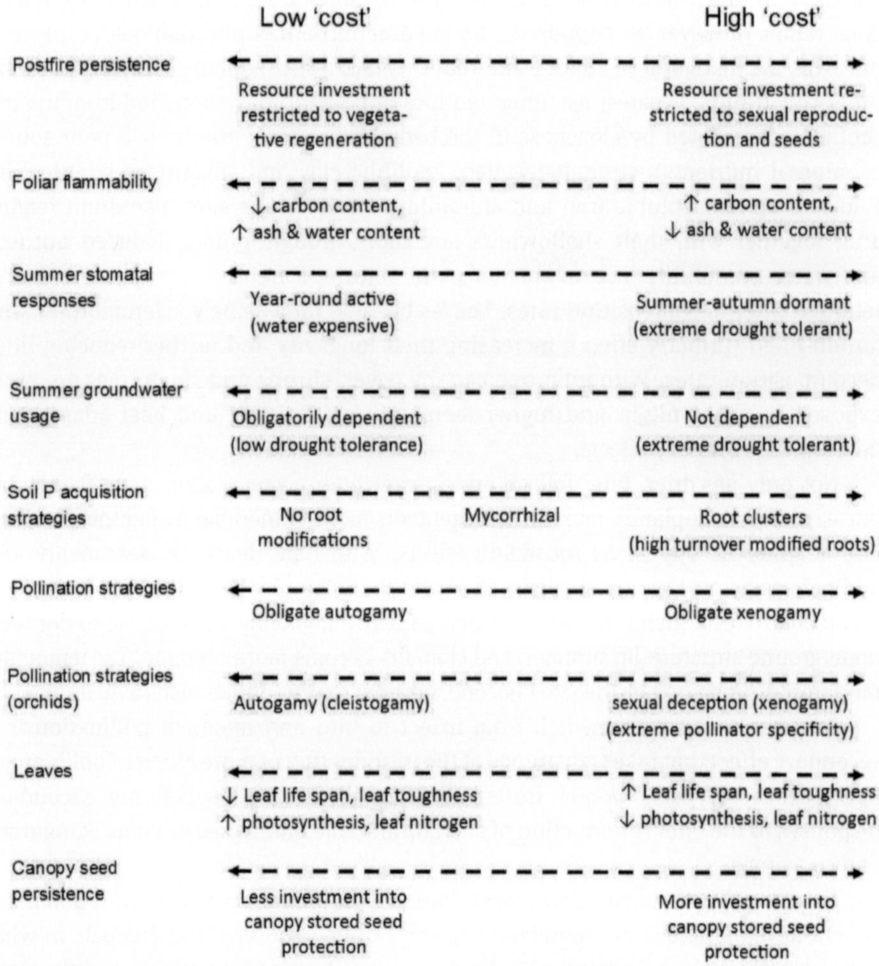

Fig. 1.2: Trait economic biospectra they relate to the SouthWest flora. The term 'cost' implies an adaptive or evolutionary trade-off between form (*i.e.* morphology) versus function as described in this book. Traits are not necessarily related, nor is the 'cost' involved similar between traits.

1.3 The SouthWest Flora 45 Million Years Ago and Now

The Early to Mid-Eocene, 45-56 Ma, was one of the hottest and wettest periods in the Earth's history and the seas and oceans rose up to 140 m above their present levels. Much of the coastline of the SouthWest was flooded for up to 300 km inland and the present Nullarbor Plain was occupied by a huge inland sea, such that several major rivers flowed eastwards into it.

One of those was the Lefroy River system that today only exists as a series of salt lakes. This was an ideal environment for sedimentation and the Eocene mudstones have proved a rich source of pollen (palynomorphs) (Itzstein-Davey, 2007). Table 1.1 gives the results of much painstaking work on identifying the plant taxa present from their pollen, compared with their current presence elsewhere in Australia.

None of the species recorded currently exists at the Lefroy River site as it is now far too dry and saline. The vegetation was a mixture of subtropical and temperate rainforest, dominated by deciduous *Nothofagus* (Fig. 1.3) and sclerophyllous woodland and shrublands dominated by eucalypts but especially many genera of Proteaceae. Indeed, the Proteaceae accounted for a mean of 34% of all pollen (when the wind-pollinated *Nothofagus* was omitted), with species richness values comparable with those in SouthWest scrub-heath today. Of the 28 rainforest taxa, 21 have become extinct in southwestern Australia (though eight still exist in the Kimberley) but 15 of the 16 hard-leaved taxa remain and have speciated greatly since then. In contrast, 26 of the 28 rainforest taxa still occur in eastern Australia, as well as all the sclerophylls, but they have speciated less.

The transformation from a predominately rainforest flora to hyperdiverse temperate woodlands and shrublands (Fig. 1.3) due to increased aridity during the Miocene and Pliocene (Byrne *et al.*, 2011) resulted in the SouthWest flora exhibiting morphological and physiological traits more suitable for seasonally dry Mediterranean-type environments (Table 1.2). About a third of Australia's total (8,500) occurs in southwestern Australia, currently the Mediterranean-climate corner, though it only occupies 5% of the Australian continent.

Fig. 1.3: (*upper left*) *Nothofagus moorei* forest in uplands of the Lamington Plateau, Queensland. Similar fire-free, temperate rainforests were the dominant vegetation type over much of Australia, including the SouthWest, during the Palaeogene. (*upper right*) Present day sclerophyllous scrub-heath vegetation of the sandplains of the SouthWest with Mt Lesueur in the background and the grasstree, *Xanthorrhoea drummondii*, in the foreground. (*lower left*) *Eucalyptus diversicolor* (karri) forest is confined to the present day wetter areas of the SouthWest. (*lower right*) outcrops of gneissic granite provide refugia for plants from fire but increased exposure to drought and summer heat.

Table 1.1: Soft- and hard-leaved plant taxa occurring in the Early to Mid-Eocene, 40-56 Ma, according to pollen samples collected from the Lefroy River system, southwestern Australia, compared with their current presence and occurrence elsewhere in Australia. Data collated by Byron Lamont from Hos (1975), Itzstein-Davey (2004), *Florabase* and Flora of NSW.

Leaf type	Modern taxon equivalent	Common name	Still present in SouthWest?	Present in Eastern Australia?
Soft-leaved (closed forest)	*Beauprea*	Protea family	no (New Caledonia)	no
	Adansonia	Sausage tree	no (Kimberley)	no
	Nothofagus (dominant)	Southern beech	no	yes
	Araucaria	Conifer	no	yes
	Agathis	Conifer	no	yes
	Dacrydium	Conifer	no	yes (Tas)
	Microcachrys	Conifer	no	yes (Tas)
	Livistona	Palm	no (Kimberley)	yes
	Ficus	Fig	no (Kimberley)	yes
	Ilex	Holly	no (Kimberley)	yes (Qld)
	Musgraveinae	Protea family	no	yes
	Anacolosa	Galonut (India)	no (*Olax* related)	yes
	Sonneratia	Mangrove	no (Kimberley)	yes
	Symphionema	Protea family	no	yes
	Lomatia	Protea family	no	yes
	Gleichenia	Scrambling fern	no	yes
	Cyathea	Tree fern	no	yes
	Eugenia/Syzygium	Myrtle	no (Kimberley)	yes
	Sapotaceae	Native plum	no (Kimberley)	yes
	Polypodiaceae	Epiphytic fern	no (Kimberley)	yes
	Lycopodium	Clubmoss	no	yes
*possibly *Nuytsia* as no other mistletoe existed then	Loranthaceae*	Mistletoe	yes	yes (not *Nuytsia*)
	Hibiscus	Hibiscus	yes (Eremean)	yes
	Dennstaedtiaceae	Bracken fern	yes	yes
	Selaginella	Clubmoss	yes (1 sp)	yes
	Podocarpus	Conifer	yes (1 sp)	yes

continued **Table 1.1:** Soft- and hard-leaved plant taxa occurring in the Early to Mid-Eocene, 40-56 Ma, according to pollen samples collected from the Lefroy River system, southwestern Australia, compared with their current presence and occurrence elsewhere in Australia. Data collated by Byron Lamont from Hos (1975), Itzstein-Davey (2004), *Florabase* and Flora of NSW.

Leaf type	Modern taxon equivalent	Common name	Still present in SouthWest?	Present in Eastern Australia?
#1 location near Denmark	*Sphagnum*	Peat moss	yes (1 sp)#	yes
	Cunoniaceae	-	yes - *Aphanopetalum*	yes
Hard- leaved (woodland, shrubland)	*Banksia*	Protea family	yes (abundant)	yes
	Adenanthos	Protea family	yes (abundant)	yes (rare)
	Hakea/Grevillea	Protea family	yes (abundant)	yes
	Lambertia	Protea family	yes	yes
	Petrophile	Protea family	yes (abundant)	yes
	Synaphea	Protea family	yes (abundant)	yes
	Xylomelum	Protea family	yes	yes
+ unidentified	Many other Proteaceae +	Protea family	yes	yes
	Eucalyptus	Myrtaceae	yes (abundant)	yes (abundant)
	Angophora	Myrtaceae	no	yes
	Other capsulate Myrtaceae	Myrtaceae	yes (abundant)	yes (abundant)
	(Allo)Casuarina	She-oak	yes (abundant)	yes
	Macrozamia	Cycad	yes	yes
	Ericaceae	Epacrid	yes (abundant)	yes (abundant)
	Dodonaea	Hopbush	yes (abundant)	yes (rare)
	Beyeria	Turpentine bush	yes (abundant)	yes (rare)

Table 1.2: Generalised shifts in Angiosperm form and function in the SouthWest flora from a predominantly rainforest environment (Cretaceous/Palaeogene) to a predominantly woodland environment (Neogene/Quaternary).

Rainforest (closed vegetation)	Woodland (open vegetation)
Drought-sensitive (mesophytes)	Drought-tolerant (xerophytes)
Leaves thin, light, moist (mesophylls)	Leaves thick, dense, dry (sclerophylls)
Fire-sensitive	Fire-tolerant
High-nutrient requiring	Low-nutrient tolerant
Salt-sensitive (glycophytes)	Salt-tolerant (halophytes)
Single-stemmed (tree)	Multi-stemmed (mallee)
Thin barked	Thick barked
Loss of leaf bases	Retention of leaf bases (e.g. *Dasypogon*)
Shoots of equal size	Long and short shoots (heteroclady) (e.g. *Allocasuarina*)
Plants large, fast growers	Plants small, slow growers
Tree ferns	Grasstrees (*Xanthorrhoea*)
Herbs aerially located	Herbs terrestrial (e.g. orchids)
Wood poorly lignified (soft)	Wood highly lignified (hard)
Growth non-seasonal	Growth seasonal
Leaves horizontal, stomates on one side (bifacial)	Leaves vertically twisted (tips pointing upwards) or erect, stomates on both sides (isobilateral)
Leaves flat	Leaves rolled at edges (ericoid) or circular in cross section (terete)
Stems with leaves	Stems with cladodes, phyllodes or phylloclades
Fruits succulent, indehiscent	Fruits woody, dehiscent (e.g. Myrtaceae)
Seeds without ant-attracting arils (elaiosomes)	Seeds with elaiosomes (e.g. legumes)
Plants palatable	Plants unpalatable
Tannins, essential oils and resins rare	Tannins, essential oils and resins abundant

1.4 A Relict Podocarp

Podocarpaceae is an ancient conifer family that had its origins 240 million years ago (Triassic) though the modern species had their origins more recently, 125 million years ago. They were widespread in Australia in the Cretaceous and Paleogene and ten species still occur here, all but two in rainforests on the east coast. The only survivor in the SouthWest is *Podocarpus drouynianus* that occurs in the jarrah forest (Fig. 1.4).

It would have been eliminated along with the other rainforest species in the drier late Neogene-Quaternary but for the possession of two special features: 1) it is a rhizomatous shrub (all but one other species are substantial trees) that survives fire and moreover, coning is stimulated by fire (Chalwell & Ladd, 2005; Ladd & Enright, 2011); and 2) it has exceptionally narrow, ribbed, glaucous and semi-erect leaves that, combined with its compact, low-lying growth form, appear to increase its drought tolerance compared with other podocarps. So *P. drouynianus* is a relict from wetter times only surviving because it is well matched to its current mediterranean climate and frequent fire.

Fig. 1.4: *Podocarpus drouynianus* (Podocarpaceae) is a rhizomatous shrub, 75 cm tall, and restricted to the higher rainfall areas of SouthWest's jarrah forests. *P. drouynianus* is uniquely fire- and drought-tolerant among podocarps that normally exist as trees in rainforest throughout the Southern Hemisphere.

1.5 Fire and Evolution of the Proteaceae

A time-based phylogeny for one of the most prominent plant families in the Southern Hemisphere, the Proteaceae, is presented in Fig. 1.5. The Proteaceae is especially well represented in the SouthWest flora with 800 extant species. The phylogeny was created by comparing parts of the DNA structure for species in all genera in the family to show their phylogenetic relationships independent of their morphological relationships (called a molecular phylogeny; see Lamont & He (2012) for more details).

The dates for seven well-chronicled fossils (preserved leaves and pollen) scattered through the geological record were compared with differences in DNA structure through the phylogeny to create a molecular clock (from left to right) for the stems of each lineage (line of descent) and thus a chronogram for the evolution of the entire family.

Whether each genus occurs in fire-prone environments is given in red, or non-fire-prone (usually rainforest) environments is given in green. The three genera in eastern Australia with species in both are given in alternating red and green. The genera were then progressively fused from right to left using the seed-storage and fire-environment criteria in separate analyses, using a technique called the Bayes Markov Chain Monte Carlo procedure, to reconstruct ancestral traits in the family. The closely related Sabiaceae, *Nelumbo* (Nelumbonaceae) and *Platanus* (Platanaceae) were used as a reference group in the chronogram for the Proteaceae.

The chronogram shows that the Proteaceae arose 115 Ma with the oldest clades appearing 94 Ma. This makes it one of the oldest flowering plant families known. The Proteaceae soon separated into two further groups that are by far the most speciose today, the subfamilies Grevilleoideae and Proteoideae. Basal lineages, such as *Placospermum*, *Agastachys/Symphionema*, *Carnarvonia* and Macadamieae, arose 80-90 Ma in rainforest and still occur there along the eastern border of Australia.

However, the Proteoideae arose in a fire-prone environment 90 Ma (Fig. 1.5) and speciated strongly once its ancestor left the rainforest, giving rise to such major sclerophyll-shrub genera as *Petrophile* (49 Ma), *Conospermum/Synaphea* (36 Ma) and *Adenanthos* (34 Ma) now accounting for 250 species in the SouthWest flora. Fire-adapted lineages arose much more slowly and fitfully in the Grevilleoideae, with the major sclerophyll shrub/tree genera, *Banksia* (61 Ma) and subgenus *Dryandra* (23 Ma), *Hakea* (18 Ma) and *Grevillea* (11 Ma) now accounting for over 500 species in the SouthWest flora.

By 81 Ma, fire-adapted traits had begun to appear. In particular, the *Petrophile* lineage evolved with the ability to hold its fruits in a woody cone until released by fire heat (serotiny). *Conospermum/Franklandia* and *Adenanthos* appeared with the ability to store single-seeded fruits in the soil until cued to germinate by smoke. A refinement at 44.5 Ma was the development of arils (elaiosomes) on the diaspores that attract ants that bury them to an ideal depth for receiving a fire heat/smoke pulse to stimulate germination. This is one of the earliest records for elaisome-bearing seeds or fruits (diaspores) known. Serotiny first appeared among the Grevilleoideae 61 Ma *via Banksia*, followed later by three other serotinous lineages. Fire-enhancing traits, such as the retention of dead leaves by *Banksia bella* (= *Dryandra pulchella*) (Fig. 1.6) first appeared 26 Ma in the *Banksia* clade. *Grevillea*, which originated much later, is the only taxon in Proteaceae with seeds (rather than fruits) bearing arils or palatable wings, and is one of the most speciose genera in the Australian flora, third only to *Acacia* and *Eucalyptus*.

All of these fire-related traits are highly correlated through time with the occurrence of fire and, coupled with their fitness benefits in the presence of fire, confirm their place as important fire adaptations in the SouthWest flora.

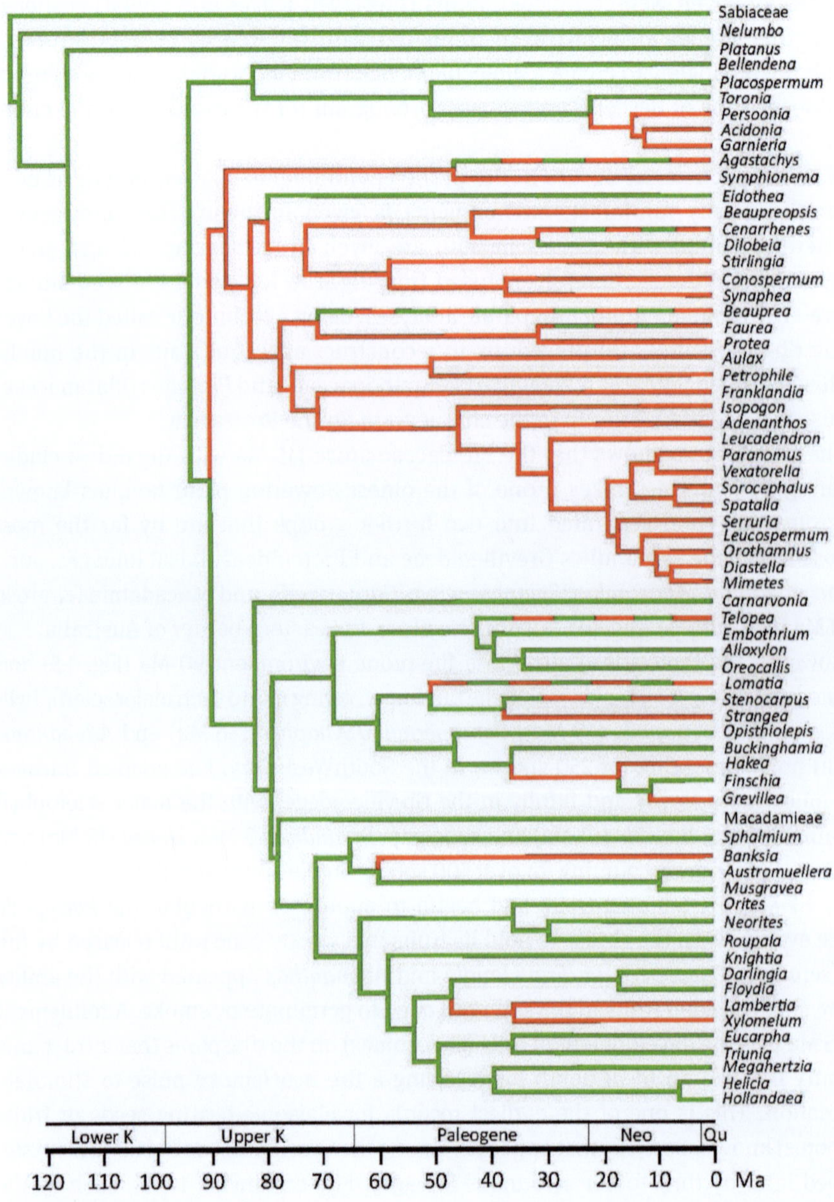

Fig. 1.5: Chronophylogeny of extant genera in the Proteaceae family. Adapted from Lamont and He (2012) based on data obtained from Sauquet *et al.* (2009). Genera and lineages that currently occur in fireprone floras are given in red, with broken lines indicating lineages with both extant rainforest and fireprone-habitat species. Genera in Macadamieae were collapsed into one lineage. K = Cretaceous, Neo = Neogene, Qu = Quaternary. SouthWest genera include *Persoonia*, *Acidonia* (endemic) *Stirlingia* (endemic), *Conospermum*, *Synaphea* (endemic), *Petrophile*, *Isopogon*, *Franklandia* (endemic), *Adenanthos*, *Strangea*, *Hakea*, *Grevillea*, *Banksia*, *Lambertia* and *Xylomelum*.

Fig. 1.6: Some species in the family Proteaceae. (*top left*) *Petrophile longifolia* (Proteaceae, subfamily Proteoideae) 30 cm tall, unlike other members of the genus bears its flowers just above the ground. (*top right*) *Conospermum stoechadis* (Proteoideae). Plant is 60 cm tall. (*middle left*) *Stirlingia latifolia* (Proteoideae) bearing male and bisexual flowers, 8 mm long. (*middle right*) *Persoonia saccata* (Persoonioideae), drupes 1 cm wide. Species with succulent fruits have rainforest affinities and are now rare in the SouthWest flora. (*lower left*) *Banksia bella* (previously *Dryandra pulchella*) (Grevilleoideae), a species that retains its dead foliage. This fire-enhancing trait first appeared 26–16 Ma (Miocene) in *Banksia* with the onset of seasonal drought and thus more frequent fire (He *et al.*, 2011). Plants are 2 m tall. *(lower right) Hakea pandanicarpa* (Grevilleoideae). Individual flowers are 1.5 cm long. Note ferruginous new growth due to the dense indumentum of orange hairs.

1.6 Floristics and Phytogeography

There are over 1,400 genera of flowering plants, 7 Gymnosperm genera and 44 fern genera currently recognised as occurring in Western Australia. Over 7,100 named taxa are now known to occur within southwestern Australia, and the actual number could be as high as 8,500. More than 250 new plant species in the SouthWest were discovered and named in 2013 alone. In 1989, 3,560 plant taxa were recognised as occurring in Western Australia. This marked increase over the last 30 years can be accounted for by improved accessibility into more remote parts of the State, an increase in the number of botanical surveys and the continual taxonomic re-examination and reclassification of the flora at the genus and species levels.

Western Australia can be divided into 26 vegetation regions (phytoregions) that are based on differences in climate, soils and physiography (Fig. 1.7). The boundaries of these regions have been incorporated into an Interim Biogeographic Regionalisation for Australia (IRBA) (Australia's Bioregions, 2014). The IRBA bioregions and sub-regions currently used are the reporting unit for assessing the status of natural ecosystems, their protection in Australia's national reserve system and are also used as the monitoring and evaluation framework for the Australian Government's current Natural Resource Management initiatives.

These phytoregions are grouped into three Botanical Provinces: the Northern Province (the Kimberley); Eremaean Province (the desert interior, Murchison, Pilbara and Gascoyne) and the SouthWest Province. An additional zone, known as the SouthWest Interzone, is considered by some to be part of the Eremaean Province.

The concept of a southwestern Australian bioregion was first developed by Ferdinand von Müller in 1877 and subsequently discussed in detail in Ludwig Diels book *The Plant Life of Western Australia South of the Tropics* in 1906. In 1944, Charles Gardner (Western Australia's Government Botanist, 1929-1960) published a paper entitled *The vegetation of Western Australia with special reference to climate and soils* that expanded upon Diels' work in documenting the State's phytogeographic regions.

John Beard, inaugural director of Kings Park and Botanic Gardens in Perth, undertook the project of mapping Western Australia's vegetation from 1965 to 1985. This involved extensive field surveys and mapping from aerial photography that further refined the main state botanical regions (Beard, 1980) (see Fig. 1.7 and Table 1.3).

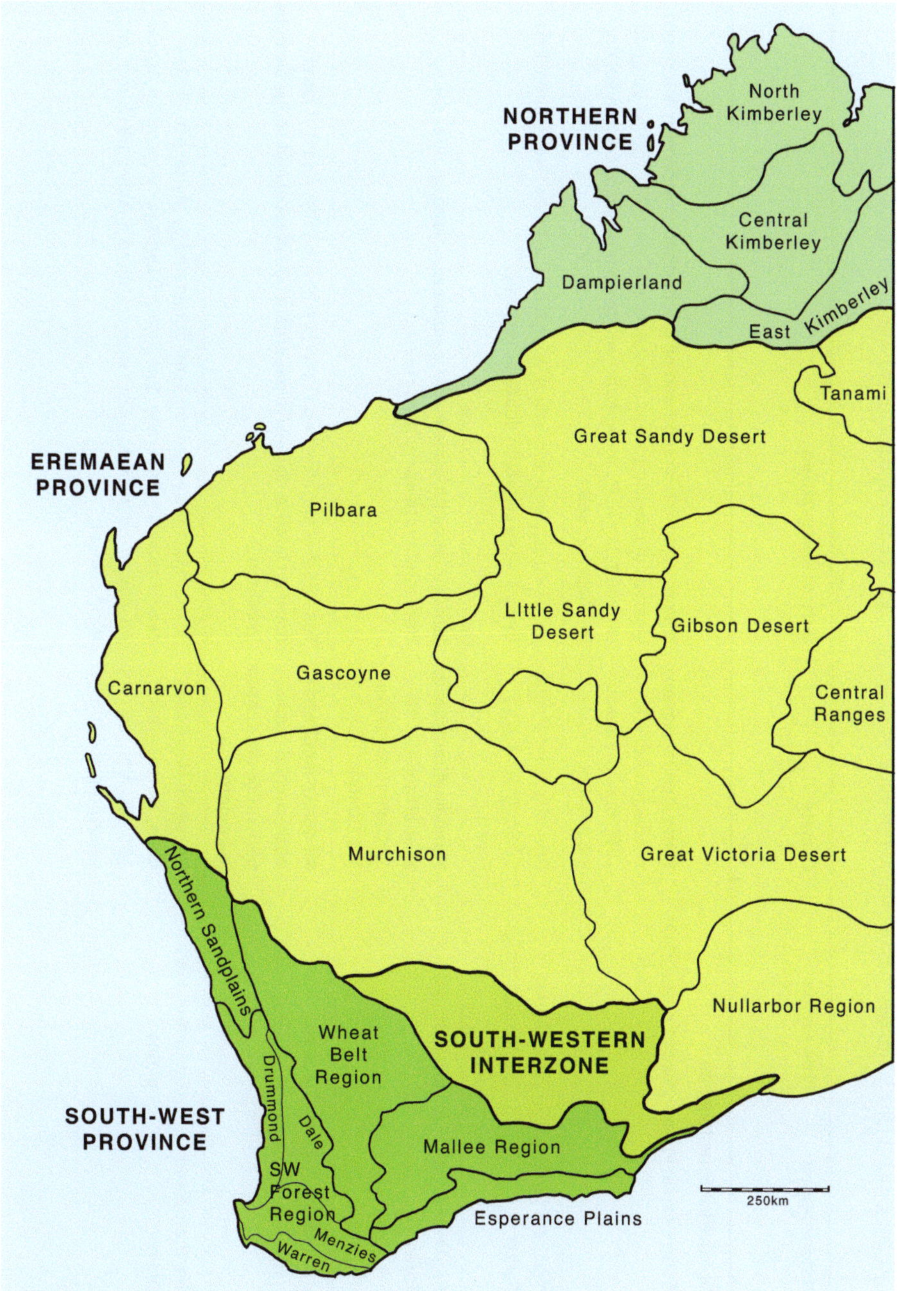

Fig. 1.7: Phytoregions of Western Australia (SouthWest coloured in dark green). See Table 2.3 for descriptions of the SouthWest phytogregions (after Beard, 1990).

Table 1.3: Phytogeographic groups of the SouthWest based on Beard (1980; 1990). See Fig. 1.8 for photographs of the main vegation types.

Botanical District or sub-district (SD)	Natural region	IRBA region	Climate	Dominant vegetation
Darling District	**SouthWest Forest Region**			
Drummond SD	Swan Coastal Plain	Swan Coastal Plain	**Warm mediterranean**; annual rainfall 600-1000 mm; dry season 5-6 months	Mainly banksia low woodland on leached sandy soils with *Melaleuca* swamps in low-lying areas. Woodlands of tuart (*Eucalyptus gomphocephala*), jarrah (*E. marginata*) and marri (*Corymbia calophylla*) on less leached soils.
Dale SD	Northern Jarrah Forest	Jarrah Forest (*Eucalyptus marginata*)	Drier than, but similar to, Southern Jarrah Forest	Jarrah forest, with some marri-wandoo (*E. wandoo*) woodlands. Relatively sparse sclerophyllous understorey.
Menzies SD	Southern Jarrah Forest	Jarrah Forest	**Warm mediterranean**; annual rainfall 600-1200 mm; dry season 5-6 months	Jarrah forest with some marri- wandoo woodland on drier soils. Understorey similar to Karri Forest.
Warren SD	Karri Forest	Warren	**Moderate mediterranean**; annual rainfall 650-1400; dry season 3-4 months	Tall forest of karri (*E. diversicolor*) on fertile soils; forest of jarrah-marri on laterite. Extensive paperbark (*Melaleuca*) and sedge swamps in valleys.
Irwin District	Northern Sandplains	Geraldton sandplains	**Dry warm Mediterranean**; annual rainfall 300-500 mm; dry season 7-8 months	Scrub-heath (kwongan) on sand-dominated by Myrtaceae, Proteaceae, Fabaceae, Ericaceae. Species-rich kwongan vegetation *Acacia-Allocasuarina* thickets inland with scattered prickly bark (*E. todtiana*) more coastal.

continued **Table 1.3:** Phytogeographic groups of the SouthWest based on Beard (1980; 1990). See Fig. 1.8 for photographs of the main vegation types.

Botanical District or sub-district (SD)	Natural region	IRBA region	Climate	Dominant vegetation
Avon District	Wheatbelt region	Avon Wheatbelt	**Dry warm mediterranean**; annual rainfall 300-650 mm; dry season 7-8 months	Woodlands of york gum (*E. loxophleba*), salmon gum (*E. salmonophloia*) and wandoo (*E. wandoo*). *Acacia-Allocasuarina* thickets.
Roe District	Mallee region	Mallee	**Dry warm mediterranean**; annual rainfall 300-500 mm, dry season 7-8 months	Mallees are small species of eucalypts that have numerous thin stems arising from a swollen rootstock (lignotuber). Mostly *E. eremophila*. Dense understorey of sclerophyllous shrubs, mainly *Melaleuca.*
Eyre District	Esperance Sandplains	Esperance sandplains	**Moderate mediterranean**; annual rainfall 500-700 mm; dry season 5-6 months	Vegetation is kwongan (scrub-heath, mallee-heath) on sand (over laterite or clay). *E. tetragona* is a common mallee.

Fig. 1.8: (*upper left*) Banksia woodland of the Swan Coastal Plain (Drummond Subdistrict), with *Banksia attenuata* and *B. menziesii* co-dominating the overstorey, with scattered *B. ilicifolia* and jarrah (*Eucalyptus marginata*, far right). *Hibbertia hypericoides* (Dilleniacee) is in flower in the low shrub layer. (*upper right*) Jarrah forest of the Darling Scarp (Dale Subdistrict), 50 km southeast of Perth. (*middle left*) tall karri (*Eucalyptus diversicolor*) forest of the Warren Subdistrict. (*middle right*) Species-rich shrubland (scrub-heath) in the northern sandplains (Irwin District) near Eneabba with 1.5-m-tall *Banksia menziesii* in flower (*left*) and *B. hookeriana* (*right*) with non-flowering *B. attenuata* in between. 113 perennial species were recorded in a 40 × 40 m plot at this site (Herath *et al.*, 2009). (*lower left*) Mixed eucalypt woodland of the Wheatbelt (Avon District) is now confined to isolated reserves and roadsides. Pictured is a salmon gum woodland (*Eucalyptus salmonophloia*). (*lower right*) Species-rich shrublands of the sandplains surrounding the Barren Range, with the iconic erect *Hakea victoria* (Proteaceae) standing 2.5 m tall in the mid-foreground.

1.7 The SouthWest as a Biodiversity Hotspot

The uniqueness of modern day SouthWest flora is well recognised and documented, and is a major biodiversity hotspot, globally and within Australia (Hopper & Gioia, 2004). The definition of a hotspot is an area that contains at least 0.5% (or 1,500) of the world's plant species as endemic and whose biota is under threat from human activity, such as land clearing (Myers *et al.*, 2000). The only Australian hotspot that meets these criteria is southwestern Australia.

The flora is both species and phylogenetically rich at both the regional and landscape levels (Sauquet *et al.*, 2000; Hopper & Gioia, 2004; Merwin *et al.*, 2012), and contains numerous endemic genera and species (Hopper & Gioia, 2004) some of which represent ancestral forms of pandemic families (e.g. *Nuytsia*; Hopper, 2010) and ancient lineages (e.g. order Dasypogonales and family Cephalotaceae). The most species-rich plant families and genera are listed in Table 1.4.

For example, the high-rainfall karri forest communities contain local endemic forest eucalypts (the tingles – *Eucalyptus brevistylis, E. jacksonii, E. guilfoylei* and the red-flowering gum, *Corymbia ficifolia*) that are considered relictual biota of past wetter climates that are restricted to the southern coast of the SouthWest (Wardell-Johnson & Coates, 1996). Another endemic relict, *Cephalotus follicularis* (Albany pitcher plant), is restricted to wetlands and fringes of bogs and drainage ditches. These forested regions have one of highest levels of orchid species richness in the SouthWest (Phillips *et al.*, 2011).

This extraordinary species richness is reflected in the multitude of morphologies and physiological processes evolved at the genus, species, community and ecosystem levels that maintain population survival and expansion in their nutrient-impoverished, seasonally dry and fire-prone landscapes. Traits may vary both along environmental gradients and among species occupying similar habitats (Lamont *et al.*, 2002; Ackerly & Cornwell, 2007), and are the foundations for understanding adaptive responses to past, present and future changing climates (Esther *et al.*, 2010).

Among the world's five Mediterranean-type climates, the SouthWest region is topographically unique in that it is a low, stable, highly weathered plateau with granite outcrops (monadnocks or inselbergs) occasionally emergent. There are a few small mountainous areas, with a maximum elevation just over 1000 m (in the Stirling Range) (Fig. 1.9). The nutrient-impoverished, seasonally dry soils of the SouthWest have resulted in a flora evolving that is predominantly sclerophyllous (literally 'hard-leaved') trees, shrubs and perennial herbs (Hopper, 2009). About 7% of the flora contains monocotyledonous geophytes (Parsons & Hopper, 2003). About 58% of the SouthWest vascular flora currently is endemic here (Mucina *et al.*, 2014). This value can be expected to rise to 75% based on the current rate of descriptions of taxa, almost all of which are endemic, and as phytogeographic methods move away from the grid system of determining occurrence that overestimates species distribution east of the SouthWest. Endemism and species richness is greatest by far in the upland areas, especially the Mt Lesueur-Eneabba region, the Stirling Range and the Barren Ranges (Hopper, 1979; Groom & Lamont, 1996a; Lamont & Connell, 1996; Beard *et al.*, 2000) (Fig. 1.9). It is these areas that best define the uniqueness of the SouthWest flora.

Fig. 1.9: (*upper left*) The Stirling Range. Bluff Knoll is the highest point in the SouthWest (1,099 m above sea level (asl)). (*upper right*): The Barren Ranges, Fitzgerald River National Park. West Mt Barren is 377 asl. (*lower left*) Mt Lesueur (313 m asl), a near-circular flat-topped mesa. (*lower right*) species-rich shrubland surrounding Mt Manypeaks (565 m asl), 35 km NE of Albany on the south coast of the SouthWest. Maximum height of shrubland species is 2.5 m. Banksias in flower are *B. coccinea* (red) and *B. attenuata* (yellow) to a maximum height of 2 m, with a white-flowered subshrub, possibly *Sphenotoma* sp. (Ericaceae), abundant.

Ironstone ranges and granite outcrops of the woodlands and arid shrublands exhibit high levels of local endemism, differing in species composition, diversity and microclimate from the surrounding landscapes (Gibson *et al.*, 2010; Schut *et al.*, 2014). Species restricted to fissures in these formations have specialised root morphologies that enable access to stored moisture sources in the underlying rock (Poot & Lambers, 2008; Poot *et al.*, 2012). Both granite outcrops and ironstone ranges may have acted as refugia during the transition to drier, seasonal cclimates of the Cenozoic Era (Gibson *et al.*, 2012; Keppel *et al.*, 2012).

Table 1.4: The ten taxa-richest families and genera in the SouthWest with number of taxa in parentheses. Data obtained from Florabase (http://florabase.dpaw.wa.gov.au/), Western Australian Herbarium's online taxa database. Data accessed 10 October 2013. *Melaleuca* was expanded in 2014 to include many other related genera not added here.

Families	Genera
Myrtaceae (1268) – Eucalypt family	*Acacia* (438) – Fabaceae
Fabaceae (1026) – Legume family	*Eucalyptus* (313) – Myrtaceae
Proteaceae (765) – Banksia family	*Leucopogon* (209) – Ericaceae
Orchidaceae (382) – Orchid family	*Stylidium* (208) – Stylidiaceae
Ericaceae (371) – Epacrid family	*Grevillea* (195) – Proteaceae
Asteraceae (290) – Daisy family	*Melaleuca* (182) – Myrtaceae
Cyperaceae (250) – Sedge family	*Banksia* (156) – Proteaceae
Stylidiaceae (217) – Triggerplant family	*Caladenia* (146) – Orchidaceae
Goodeniaceae (204) – Wing-petalled family	*Gastrolobium* (111) – Fabaceae
Malvaceae (184) – Mallow family	*Hakea* (100) – Proteaceae
	Baeckea (100) – Myrtaceae

2 Fire Adaptations

Fire is a natural component of the Australian landscape and the vegetation has had to adapt to fire-prone environments since at least the Cretaceous (90 Ma). Recent studies, mostly centred on Australia, have confirmed the importance of fire as a major global and regional disturbance shaping the world's flora (Bond & Scott, 2010; Burrows, 2013; Crisp *et al.*, 2011; He *et al.*, 2011; He *et al.*, 2012; Lamont & He, 2012; Lamont *et al.*, 2013). Traits such as resprouting, heat-stimulated release of canopy-stored seeds (serotiny), heat- and smoke-cued germination and post-fire stimulated flowering are common features of today's SouthWest flora. These traits appear in many phylogenetically-independent plant families occupying the world's fire-prone environments (Verdu *et al.*, 2007; Crisp *et al.*, 2011; He *et al.*, 2011; Litsios *et al.*, 2014). Two of the SouthWest's most iconic genera, *Banksia* (Proteaceae) and *Eucalyptus* (Myrtaceae), occur in fire-prone ecosystems throughout Australia and first appeared about 61 Ma (Burrows *et al.*, 2011; He *et al.*, 2011; Fig. 1.1), with serotiny and retention of highly-combustible, dead flowers considered ancestral among banksias (He *et al.*, 2011). Seed storage in general among Proteaceae has an even longer history stretching back to 88 Ma (Chapter 1; Lamont & He, 2012).

Fire creates an ideal opportunity for vegetative recovery because between-plant competition is reduced and resource availability is enhanced (Fig. 2.1). More light is available (reduced canopy cover) post-fire and the resulting ashbed creates a flush of inorganic nutrients available for plant growth. For modern day fire-prone floras, functional traits are considered *fire-adaptive* if the trait enhances species persistence post-fire (Pausas *et al.*, 2004) and is related to the system's fire regime (Keeley *et al.*, 2011). It is considered a *fire adaptation* if, in addition to enhancing species fitness, the evolution of the trait is synchronized with fire-prone habitats through time (Lamont & He, 2012). Taxa can be defined as fire-killed (relying exclusively on recruiting seedlings after a fire) or resprouters (recovering by way of pre-existing protected buds) that can regrow from underground structures (e.g. lignotubers, rhizomes) (Fig. 2.2) or from buds arising from a main stem or branch (Bell, 2001).

This chapter discusses strategies employed by the SouthWest flora to utilise fire to its advantage, whether that be by promoting post-fire flowering, or trade-offs involving resource allocation between sexual reproduction and vegetative (resprouts and regrowth) structures. Not all species in fire-prone environments have life histories that have been shaped by fire. Species that do not recover from fire are rare in the SouthWest flora, and are confined to refugia that are unlikely to experience fire, such as granite outcrops, habitats that are permanently wet or have a high degree of succulence.

We can recognise five categories of responses to fire that are exhibited by the SouthWest flora, with many taxa displaying more than one type of response, especially at different stages in their life cycle. For example, the grasstree *Xanthorrhoea preissii* has fire-enhancing foliage and fire-dependent flowering (Figs. 2.1; 2.3) (Lamont *et al.*, 2004). A summary of fire-related traits and associated species is provided in Table 2.1.

Fig. 2.1: (*upper*) *Banksia hookeriana* (Proteaceae), a fire-killed species near Eneabba, southwestern Australia. Photos are of the same plant, 2 m tall, during (*left*) and after (*right*) a burn. Note that although the plant has been killed, neither the follicle-(fruit) bearing cones nor the stems were incinerated. (*lower*) Rejuvenation of a *Banksia-Eucalyptus* woodland on Perth's Swan Coastal Plain after a summer fire. Photographs represent the same site 8 days (*left*) and 8 months (*right*) after the fire. Note that the youngest foliage of the grasstree *Xanthorrhoea preissii* was not burnt. Trunks (caudexes) are about 1 m tall. Eight months post-fire the *Eucalyptus* and *Banksia* trees are starting to produce new growth from the main branches (known as epicormic regrowth), and *Xanthorrhoea* has produced long flowering spikes. Invasive grasses (wild oats, *Avena* spp., and veldt grass, *Ehrharta calycina*; Poaceae) have also taken advantage of the open spaces created by the fire.

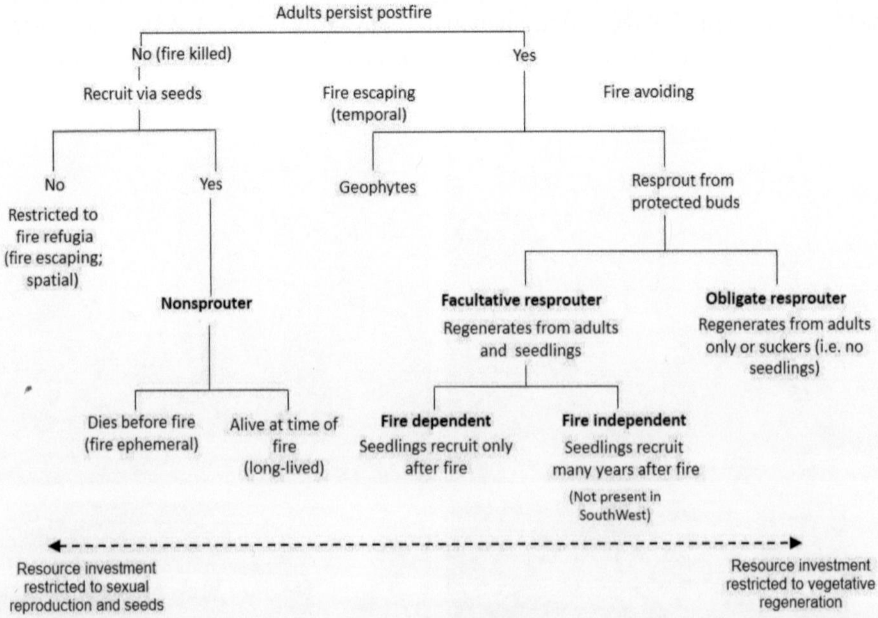

Fig. 2.2: Fire response strategies of SouthWest flora based on adult and seedling persistence post-fire. The fire-independent facultative resprouter strategy is unknown in the SouthWest flora, but a feature of the Californian chaparral (Mediterranean-type shrubland).

Table 2.1: Summary of fire adaptive features in the SouthWest flora with generic examples and notes on limitations and alternative interpretations of their function.

Category	Example	Exhibited by	Why considered a fire adaptation	Limitations to benefits	Alternative interpretations
Fire-enhancing, vegetative	Rapid evolution of heat from essential oil-bearing leaves	*Eucalyptus*	Fire perpetuated, so that benefits obtained. E.g. nutrient release, death of parasites, germination enhanced, extra light	Increases meristem damage; further reduces biomass and productivity (at least initially)	Essential oils allelopathic; repel herbivores (distasteful), reduce transpiration
	High flammability of leaves due to low nutrient content	*Eucalyptus, Xanthorrhoea, Banksia*	Fire perpetuated (as above)	As above	Result of very infertile soils and hence low nutrient requirement
	Rapid evolution of heat from extensive litter and dry unshed shoots, especially during summer drought (when shoots may die off)	Grasses, bark, twigs, leaves – especially in dry *Eucalyptus* forests and woodlands	Fire perpetuated and intensified (as above); flames carried into crowns of trees	Death of fire-intolerant species; reduction of post-fire biomass; loss of organic matter; root damage	Summer death of annuals; leaf and branch drop a drought adaptation (decreases plant transpiration); litter reduces soil evaporation; slow decay of nutrients means slow release of nutrients into system
Fire-enhancing, reproductive	Ignition of extrafloral packing around fruits	*Banksia*	Fruits require heat to open them to release seeds	Seed heat damaged or killed	Insulating material against over-heating, desiccation
Fire-dependent, vegetative	Resumption of vigorous vegetative growth and increased productivity after fire	*Daviesia, Petrophile, Conostephium, Stirlingia, Eucalyptus*	Unburnt plants become unproductive, disease-ridden	Plants rarely die in absence of fire; net effect of accelerated growth may have a negative effect on severely damaged plants; increased insect damage and grazing on new shoots; effect is short-term; vigorous growth will also occur in absence of fire	Similar regrowth may occur in response to other factors other than fire – herbivory, shoot dieback

continued **Table 2.1:** Summary of fire adaptive features in the SouthWest flora with generic examples and notes on limitations and alternative interpretations of their function.

Category	Example	Exhibited by	Why considered a fire adaptation	Limitations to benefits	Alternative interpretations
Fire-dependent, reproductive	Initiation and development of inflorescences is stimulated in 1st year after fire	*Xanthorrhoea*, *Kingia*, *Pyrorchis*, *Anigozanthos*	Rarely flower in the absence of fire; rely on increased availability of nutrients post-fire	Do not always flower after fire, and some flower without fire	General survival response to reduction in leaf biomass
	Production of inflorescences enhanced in 2nd to 4th year after fire	Many fire-tolerant shrubs; common in *Verticordia*, *Stirlingia*	Much more floriferous than in absence of fire (though fire not essential)	Reduced flowering with frequent fires; energy invested in reproductive rather than vegetative growth	Normal patterns of life cycle, whatever the cause of germination or growth spurt; general survival response
	Seeds already fallen prior to fire, require heat to crack hard seed coat to enable germination to occur	Most *Acacia* and legumes (*Hovea*, *Jacksonia*)	Species dominate after fire, gradually becoming rare, or even zero. Such species can be opportunists and are secondary successors post-fire	Sudden death of seedlings established after spring fire (due to summer drought); frequent fires result in young plants never reaching maturity; some seeds germinate in absence of fire	Hard seed coat prevents insect attack; ensures seed longevity; prevents desiccation
	High rates of germination and seedling establishment in litter microsites created by fire	*Pericalymma*, *Hypocalymma*, *Geleznowia*, *Calytrix*, *Verticordia*	Germinability can increase significantly in some species post-fire. Chemicals derived from smoke, charred wood promote germination	Germination will only occur when fire occurs. May depend on intensity of fire; other factors important as well – e.g. moisture availability. Germination may also occur outside the litter microsites (and may survive better)	
	Formation of daughter corms stimulated	*Drosera*, *Caladenia*	Enables clonal spread in the absence of viable seeds	Frequent fires may deplete reserves, plants die	General response to leaf reduction or nutrient amendment to soil, independent of fire

continued **Table 2.1:** Summary of fire adaptive features in the SouthWest flora with generic examples and notes on limitations and alternative interpretations of their function.

Category	Example	Exhibited by	Why considered a fire adaptation	Limitations to benefits	Alternative interpretations
	Matured seeds stored on plant (serotiny) released from woody fruits after fire	*Leptospermum, Banksia, Allocasuarina, Hakea*	Enables seeds to be released into a nutrient-enriched ash bed; protects seed from heat of fire	Seeds may char depending on fire intensity; seeds will be released in between fires if plant or stem dies	Seed protection from pathogens, granivores and desiccation
Fire-resistant, vegetative	Thick, corky bark that resists burning and protects vascular tissue (fire-avoiding)	*Banksia, Melaleuca, Eucalyptus*	Cambium still active after fire; epicormic buds begin to shoot	Minimal thickness required (older trees more resistant than young); cracks or scars in bark may allow entry of flames to wood; no bark completely fire resistant; some species shed their bark annually	Thick bark prevents entry by borers; protects vascular tissue from summer heat; minimises water loss from stem
	Compact, persistent leaf bases act as thick bark that prevents damage to true stem (fire-avoiding)	*Xanthorrhoea, Kingia, Macrozamia*	Leaf bases still intact after fire having acted as barrier to entry by flames, and these species are among the first to recover from fire	Leaf bases deteriorate over time; fissures develop with age and allow entry of flames to stem	Prevents entry of borers into soft stem which must survive for hundreds of years; minimises water loss, summer heat damage
Fire-resistant, reproductive	Large woody and corky fruits undamaged by fire and open to release seeds only after fire	*Hakea, Banksia, Xylomelum, Lambertia, Calothamnus, Eucalyptus, Melaleuca, Allocasuarina*	Seeds insulated from heat by fruit and are only shed onto open ground once fire has receded – taking early advantage of the colonising situation with minimal seed loss	Fruits may not mature for at least 2-3 years; some fruits may open in absence of fire, some not opening even after fire; if shed well before wet season, seeds likely to be eaten or killed by summer heat; no seeds available in non-fire years	Protection of highly nutritious seeds from insect larvae, parrots, heat (sun and fire) and desiccation. Fruits will open regardless of cause of their drying out or stem death. Delay in shedding ensures only strong winds disseminate seed. Fruits accumulate nutrients that are translocated to the seed as the fruit matures

continued **Table 2.1:** Summary of fire adaptive features in the SouthWest flora with generic examples and notes on limitations and alternative interpretations of their function.

Category	Example	Exhibited by	Why considered a fire adaptation	Limitations to benefits	Alternative interpretations
	Extrafloral packing insulates fruit and seed from full heat of fire, so damage minimal	*Banksia* including subgenus *Dryandra*	Seeds survive fire	Not as fire resistant as cork and wood; may ignite and raise temperature	Protects seeds from desiccation, predators; non-adaptive remnant of aborted flowers and redundant bracts
	High rates of germination and seedling establishment in ash-bed created by fire	*Pericalymma*, *Hypocalymma*, *Geleznowia*, *Calytrix*, *Verticordia*	Germinability can increase significantly in some species post-fire. Chemicals derived from smoke and charred wood promote germination	Germination will only occur when fire occurs, depending on fire intensity, and soil moisture availability. Germination may also occur outside the ash-bed, where no debris occurs.	
	Formation of daughter corms stimulated	*Drosera*	Enables clonal spread in the absence of viable seed	Frequent fires may deplete reserves, plants die	General response to leaf reduction or nutrient amendment to soil. Independent of fire
	Matured seed released from woody fruits after fire	*Leptospermum*, *Banksia*, *Allocasuarina*, *Hakea*	Enables seeds to be released into a nutrient-enriched ash bed; protects seed from heat of fire	Seeds may overheat depending on fire intensity; seeds will be released inbetween fires if plant or stem dies	Seed protection from pathogens, granivores and desiccation
Fire-avoiding, vegetative	Rapid regrowth from sunken epicormic (accessory) buds on undamaged stems	*Allocasuarina*, *Banksia*, *Eucalyptus*, *Nuytsia*, *Xanthorrhoea*	Enables rapid onset of shoot growth whereas surface buds killed	Epicormic shoots weak, give small-crown plant; young plants killed by fire	Enables rapid recovery from severe drought, natural pruning, dieback from root damage, flooding, pathogens

continued **Table 2.1:** Summary of fire adaptive features in the SouthWest flora with generic examples and notes on limitations and alternative interpretations of their function.

Category	Example	Exhibited by	Why considered a fire adaptation	Limitations to benefits	Alternative interpretations
	Rapid regrowth from underground, sheathed buds and unburnt portions of young leaves supported by rhizomes, bulbs, corms, tubers or caudexes	Rhizomes : *Schoenus, Lepidosperma, Lyginia, Desmocladus, Lomandra* Corms : *Stylidium, Drosera*. Caudexes: *Xanthorrhoea* Tubers: *Thysanotus*, all terrestrial orchids	Enables rapid onset of shoot growth and commencement of photosynthesis even though fire-exposed leaves killed, most of these species aestivate (summer dormant), leaving highly flammable material (fire-enhancing) on the soil surface during summer	Frequent fires may reduce underground stem growth and hence vigour of buds and flowers and, since most of these species are poor seed-setters, could finally eliminate them from the community.	Enables rapid recovery from severe drought, grazing, root damage, flooding or waterlogging, pathogens or sudden submergement by soil, rocks or litter
	Apical and axillary buds, surrounded by moist, compact leaf bases on raised caudexes	*Xanthorrhoea, Kingia*	Rapid onset of vegetative (and flowering) growth from these insulated buds after fire	Meristematic tissue reduced by frequent fires until plant finally dies	General protective strategy against desiccation, overheating and predators
	Rapid regrowth from ground-level buds on stolons	*Stylidium, Scaevola, Banksia*	Buds escape heat of fire	Not as well protected as below ground or sheathed buds; close vicinity of leaves and litter increases heat of fire	Stoloniferous habitat related to wind-exposure, nutrient deficiency, drought, grazing-avoidance and general colonising ability
	Rapid regrowth from reservoir of ground-level buds on woody, swollen lignotuber or softer rootstock	*Banksia, Hibbertia, Acacia, Eucalyptus*	Rapid onset of regrowth from lignotuberous buds even though shoots may be persistently killed by fire. Buds escape heat of fire.	Ability to recover from lignotubers often decreases with age once maturity reached; plants may still be killed by severe or frequent fires. Dormant bud reservoir can become exhausted by frequent fires.	A means of protecting dormant buds and enabling rapid and strong recovery from dieback due to drought, grazing, pathogens, and other forms of pruning, rather than a pre-adaptation to them

continued **Table 2.1:** Summary of fire adaptive features in the SouthWest flora with generic examples and notes on limitations and alternative interpretations of their function.

Category	Example	Exhibited by	Why considered a fire adaptation	Limitations to benefits	Alternative interpretations
	Initiation and elongation of adventitious buds on lateral roots (suckers)	*Grevillea, Acacia, Stirlingia, Banksia elegans*	Enables vigorous recovery and stimulates production of more individuals	Depletion of food reserves with frequent fires; suckers not as sturdy as parent	Enables rapid recovery when shoot or buds of parent damaged by heat, grazing, drought, diseases, as well as fire
Fire-avoiding, reproductive	Elaiosome-bearing seeds, carried underground by ants, germinate in the first wet season following fire	Many legumes (too numerous to mention e.g. *Bossiaea, Templetonia*) and others (e.*g. Goodenia*)	Seeds inside nest are protected from heat of fire and are at an ideal depth for germination	Seed buried too deeply; localisation of seeds lead to increase competition and and reduced establishment; seeds too shallow and killed	Burial protects seeds from surface predators and desiccation; increases longevity; increases contact with soil water for germination and establishing general dispersal attractant, especially for birds; increases contact with humus, nutrients
	Sharp-pointed grass fruits, buried by twisting of their hygroscopic awns, germinate the first wet season after fire	*Themeda*	The percentage of seeds surviving fire is much greater than many other grasses	Twisting does not always succeed in burying seed; rare trait even among grasses	Burial protects seed and facilitates germination as above

continued **Table 2.1:** Summary of fire adaptive features in the SouthWest flora with generic examples and notes on limitations and alternative interpretations of their function.

Category	Example	Exhibited by	Why considered a fire adaptation	Limitations to benefits	Alternative interpretations
	Flowers and fruits borne at, near or below soil surface such that heat damage nil or negligible	*Alexgeorgea,* prostrate *Grevillea, Banksia* species	Effects of heat from fires decreases as one moves from well above to below ground level	Seeds poorly dispersed; reduced pollination or need for specialised pollinators	Position of flowers outcome of growth habitat
	Seeds set by annuals before fire germinate in the winter following fire	Annuals	Species rare or unseen in the absence of fire	Spring fires prevent seed set; seeds not always fire-resistant; source of seeds in burnt areas may be from adjacent unburnt areas	Annuals are opportunists (colonisers) that proliferate in response to any reduction in competition by drought, flood, intense grazing, disease, new soil deposits, as well as by fire
	Seeds from unburnt areas deposited on burnt surface by birds or wind		Though virtually absent from site before fire, these species establish well after it	Stock plants destroyed by fire; dropped seeds eaten, desiccate before rains	A general dissemination strategy and more likely to be dropped on unburnt ground
Fire-intolerant, vegetative	Plants killed	Many genera	Many of these species have short life cycle and may die in the absence of fire	Frequent fires may lead to young plants never reaching maturity	Species are in no way adapted to fire as evidenced by their premature death

2.1 Fire-enhancement

Fire-enhancing species have an unusually high level of combustibility. Depending on the species, this may be due to an abundance of essential oils (universal in the Myrtaceae and Rutaceae), low ash, low water or high cellulose contents. Species that retain their dead leaves (e.g. *Xanthorrhoea preissii*) or retain their dead flowers (e.g. *Banksia baueri*) (Fig. 2.3) make the plant more flammable. This category is an indirect adaptation to fire, involving perhaps destruction of aerial pathogens and herbivores, enhanced return of nutrients to the soil (especially phosphorus, calcium and potassium) or destruction of non-fire-adapted competitors. To be adaptive, some other aspect of the species biology has to be fire-resistant, such as the presence of woody fruits that give it an advantage over its competitors.

2.2 Fire-dependence

Fire-dependent species are those species that require the passage of fire to grow, germinate or flower. Some *Banksia* species need the heat of a fire to stimulate fruit opening, and possess flammable infructescences to ensure that this occurs (Lamont & Cowling, 1984; Enright & Lamont, 1989). Fire-dependency includes species whose growth and flowering are rejuvenated post-fire (Fig. 2.4) and counters the effects of senescence.

Some resprouting species only flower for one to two years after fire so display obligate fire-stimulated flowering (e.g. some rosette *Drosera* species) (Lamont & Downes, 2011). Obligate and facultative fire-stimulated flowering species are scattered through the family Haemodoraceae (Hopper *et al.*, 2009; Lamont & Downes, 2011), with obligate fire-stimulated flowering characteristic of *Haemodorum* and *Tribonanthes*, while *Macropidia fuliginosa* and *Anigozanthos pulcherrimus* are facultative (Lamont & Runciman, 1993).

Fire ephemerals are a functional group of short-lived plants that germinate only after the passage of fire, growing rapidly and reaching reproductive maturity before the upcoming dry summer. For most of the time they exist only as a soil seedbank. These species can be found within a number of families including the Gyrostemonaceae, Apiaceae, Malvaceae, Solanaceae and Poaceae. Freshly collected seeds do not germinate in response to smoke and/or heat pulses. Seeds do, however, germinate in response to smoke in autumn after a period of burial, *i.e.* they have an after-ripening requirement and remain dormant until the next fire (Baker *et al.*, 2005a; 2005b).

Fig. 2.3: Fire enhancement. (*upper left*) *Xanthorrhoea preissii* (Xanthorrhoeaceae) an arborescent grasstree 1.5 m tall. Note the retention of dead leaves (produced since the last fire) hanging down to form a skirt. When ignited they may reach > 1000°C. (*upper right*) *Banksia baueri* (Proteaceae) cones, 15-20 cm tall. The retention of dead florets (individual flowers) by this species allows the cone to burn more intensely. The resulting heat ensures that the follicles (hidden under the florets) open in response to fire. This allows the seeds to be released later and blown into nutrient-enriched litter microsites or sand patches where they are lightly covered by debris and soil. (*lower left*) Fire-dependent and avoiding *Calothamnus graniticus* (Myrtaceae), 50 cm tall, regrowing from a below ground lignotuber after a fire. *(lower right) Banksia hookeriana* (Proteaceae), 1.7 m tall, a fire-killed species, releases its seeds that have been stored within woody fruits in response to the passage of fire. These seeds are blown onto the ground and accumulate in the post-fire debris (litter microsites). The ash derived from the burnt vegetation provides a valuable source of nutrients for post-fire seedling growth.

Fig. 2.4: Fire dependence. (*upper*) *Nuytsia floribunda* (Loranthaceae), 3 m tall, in the first flowering season after a wildfire (November-January) (*upper left*) compared with the flowering season 3 years later (*upper right*). This species survives fire and produces new growth from protected buds hidden within the stem. Flower production occurs 10 months after a fire, followed by the formation of new foliage (called protanthy) (Lamont & Downes, 2011). It is an example of facultative fire-stimulated flowering. (*lower left*) After a summer-autumn fire *Stirlingia latifolia* (Proteaceae) produces numerous new shoots arising from an underground lignotuber, followed by a profuse production of panicles during the spring (seranthy). (*lower right*) *Pyrorchis nigricans* (Orchidaceae) is entirely dependent on fire for flowering (obligate), producing foliage before flowering (seranthy) (photo taken by S. Spencer). Obligate fire-stimulated flowering occurs in many SouthWest orchid species (Lamont & Downes, 2011).

2.3 Fire-resistance

Fire-resistant species are those that do not ignite when subjected to fire heat because they have a high moisture content, or a smooth, dense, non-flammable surface (Fig. 2.5). The same species may be classified as fire-resistant (e.g. *Corymbia calophylla*, because of its bark) or fire-enhancing (as its leaves contain flammable essential oils) depending on the plant trait considered. Bark thickness is an important determinant of fire-resistance in species that have the ability to resprout epicormically post-fire (Lawes *et al.*, 2011), sometimes in combination with bark moisture content.

Fig. 2.5: (*top row*) Bark that resists burning and protects the underlying vascular and meristematic tissues. From left to right the species are *Banksia menziesii* (Proteaceae), *Corymbia calophylla* (Myrtaceae) and *Melaleuca preissiana* (Myrtaceae). (*lower left*) Seeds of *Acacia cyclops* (Fabaceae) in the process of being released from their pod. The seed, 5 mm long, is protected by a hard brown-black testa. The contorted red aril at the base of the seed is an attractant (= food source) for its animal dispersal agent; in this case it is consumed by granivorous birds (ant-dispersed species have cream-coloured arils). (*lower middle*) Large woody follicles of *Hakea platysperma* (Proteaceae), 6 cm in diameter. The thick woody fruit wall has a low water content but high cellulose/lignin content and hence provides excellent heat resistance, and also protects the seeds from granivores. (*lower right*) Coastal vegetation consists of many species that can be classified as succulent or semi-succulent (fleshy) that do not burn readily in response to fire, so these vegetation types usually fail to carry fire.

Fruits and seeds that resist burning are common within the SouthWest flora. Many dominant genera within the family Proteaceae (e.g. *Banksia, Hakea*) possess woody fruits that insulate their heat-sensitive seeds from fire (see Chapter 11).

The hard, cutinised seed coats of legumes (Fabaceae family) protect the seeds from digestion by granivores (seed-eaters) and prevent the embryo from drying out. Many legume species are ant-dispersed *via* their aril (elaiosome) and are stored in the soil seedbank. Heat resulting from fire is required to crack the thick cuticle or dislodge the strophiolar plug in the seed coat allowing the embryo to imbibe water and germinate.

Species with high foliar moisture content, e.g. succulent and semi-succulent species, are a minor component of the SouthWest flora, commonly occupying coastal dunes, estuaries and saline soils that are not fire-prone, often in the more arid parts of Western Australia.

2.4 Fire-avoidance

Fire-avoiding species are those that have strategies to escape the heat of a fire. One way is to be located on a granite outcrop where fire cannot be supported (Fig. 2.6) e.g. *Cheilanthes* and other rock ferns. Another is to avoid the fire season, as do annuals (therophytes) such as the everlasting daisy *Rhodanthe,* fire-escaping geophytes such as *Drosera* (Pate & Dixon, 1982), summer-dormant Monocotyledons such as occur in the Cyperaceae, Restionaceae, Anarthriaceae, Haemodoraceae, Orchidaceae (Pate & Dixon, 1982; Parsons & Hopper, 2003; Shane *et al.*, 2009) and fire ephemerals such as *Tersonia cyathiflora* (Gyrostemonaceae).

Species that insulate their seeds from fire, either indirectly by having a soil-stored seed bank, or by protecting their seeds *via* thick-walled fruits and extra-floral packing are also fire-avoiders. This includes species that have flowers and fruits borne near or below the soil surface (Fig. 2.6) such that heat damage is nil or negligible (e.g. *Alexgeorgea* (Restionaceae) with underground fruits).

Others have fire-resistant parts that insulate delicate tissues from heat damage. Soon after a fire, such species commence rapid regrowth from hidden buds, insulated from the flames by the bark or soil, hence 'avoiding' the heat of the fire (Fig. 2.3; 2.4) (Clarke *et al.*, 2013). This is by far the most common strategy in the dry sclerophyll vegetation of the SouthWest. New vegetative growth can be stimulated a few days after the fire, and flowering may occur within the year, depending on the species (Lamont & Downes, 2011).

Fig. 2.6: Fire-avoidance. (*upper left*) Granite outcrops provide a refuge for fire-sensitive species. (*upper right*) Many droseras (sundews) only exist as stem tubers or corms during the dry summer-autumn that enable the dormant plant to avoid the heat of a fire during the fire-prone season, producing foliage again when the winter rains occur (shown is *Drosera erythrorhiza* (Droseraceae), 5 cm diameter). (*lower left*) *Banksia goodii* (Proteaceae), a declared rare species. This species is one of six prostrate (ground-hugging) banksias; although banksias occur widely in eastern Australia none has a prostrate growth form. The woody follicles protect two winged seeds. The old highly-combustible florets are retained ensuring that sufficient heat is received to melt the resin that seals the two valves of the follicles. The stems (rhizome) are completely below ground and insulated from flame heat by the sandy substrate. Cones 12 cm tall. (*lower right*) For grasstrees (*Xanthorrhoea* species) the apical bud is sunken and protected by the developing leaves that have a high moisture content, thus avoiding the heat of a fire (Lamont *et al.*, 2004). Fire promotes leaf production and most of the living foliage burnt during a fire is replaced within ten months.

2.4.1 Post-fire Persistence *via* Resprouting

Fires either kill plants or the adults recover by regrowing post-fire from numerous protected buds (Pausas & Keeley, 2014). Resprouters can survive many fire cycles and may live for hundreds of years. The ability to resprout accounts for 65–80% of the flora in SouthWest ecosystems (Lamont & Markey, 1995; Groom & Lamont, 1996a; Enright *et al.*, 2007), with some species losing their ability to produce viable seeds altogether.

Resprouting takes on many forms in the SouthWest angiosperm flora. Woodland and forest trees (*Eucalyptus* and *Banksia*) tend to resprout epicormically as they are more likely to experience crown fires. Where these genera are exposed to more frequent fires (e.g. northern sandplains; Fig. 1.7) species dominate that are lignotuberous shrubs and/or possess rhizomes or root suckers. Individual resprouters, such as *Banksia attenuata* (epicormic/lignotuberous), *Eucalyptus loxophleba* (epicormic), *Acacia saligna* (root suckerer), *Amphipogon turbinatus* (rhizomatous) and *Hakea prostrata* (epicormic/lignotuberous), are the most widespread species in the SouthWest, testament to their great habitat and fire-regime tolerances.

Persisting post-fire *via* resprouts is a function of vegetative bud bank protection, their location, and resources available to sustain post-fire growth (Clarke *et al.*, 2013), and is dependent on the fire intensity, severity, frequency and season. Regrowth mostly occurs from multi-budded underground structures (lignotubers or rhizomes) or aerially *via* accessory (epicormic) buds on the major branches (Fig. 2.7). Less common in the SouthWest flora are apical resprouters, e.g. *Xanthorrhoea* (Lamont *et al.*, 2004).

2.4.2 Lignotubers

For woody shrub and tree species that resprout basally post-fire, regrowth occurs from a lignotuber—a woody structure developed from the axils of the cotyledons that occurs just below the soil surface and is a source of dormant accessory buds. The lignotuber stores starch that is then used to assist in the formation of new vegetative growth after a fire, and sometimes in response to intense herbivory or drought (Clarke *et al.*, 2013). Both the soil and bark provide the buds with insulating protection from the heat of the fire. Most above-ground material is killed or consumed in a fire. Highly intense fires, or too frequent fires, favour lignotuberous regrowth /root or shoot suckering rather than epicormic regrowth as they destroy the above-ground shoots.

There is great variation among resprouting eucalypts (*Eucalyptus* and *Corymbia*) as they can be lignotuberous, epicormic or a combination of the two (Nicolle, 2006). This has implications for plant traits such as stature, bark thickness and on-plant seed storage. Lignotuberous eucalypts (mallees) are relatively short in stature and multi-stemmed with thin bark and better protected canopy-stored seeds in contrast to the tall epicormic species.

Resprouting only improves fitness in fire-prone environments if so few seeds are produced/stored that self-replacement is unlikely should the parent die, or if post-fire conditions for seedling recruitment are unfavourable or unpredictable. In this case, population survival is enhanced by parent recovering in response to fire, rather than relying exclusively on the recruitment of seedlings (Lamont *et al.*, 2011).

Fig. 2.7: New growth arising epicormically from the main stem of *Melaleuca preissiana* (Myrtaceae) (*upper left*) and *Banksia attenuata* (*upper right*) following a fire. (*middle left*) Schematic drawing through a young eucalypt stem, 5 mm wide, showing a series of accessory buds in a concealed cavity running vertically beneath the original primary surface bud that shoot progressively in response to damage of the primary stem. (*middle right*) Epicormic bud strand running radially through bark of a *Eucalyptus rudis* tree as viewed under a microscope. Largest, outermost bud 2 mm long. (*lower left*) *Verticordia grandis* (Myrtaceae) (crown 100 cm wide) recovering from a lignotuber with prominent fire-stimulated flowering. (*lower right*) *Banksia candolleana* (Proteaceae) producing many new shoots from short rhizomes (here 25 cm long) 6 months after a fire. Knowing the rhizomes increase in length 5–10 cm between fires that occur on average at 15-year intervals and the widest clone exceeds a diameter of 10 m, it can be shown that this species must reach an age of about 1 200 years (Merwin *et al.*, 2012). Image of epicormic bud strands provided by Geoff Burrows.

2.4.3 Epicormic Growth

Epicormic growth arises from latent or dormant buds underneath the bark that persist for an indefinite number of years. These buds can be activated at any time during the life of the tree in response to various stimuli, including fire and severe drought, enabling rapid crown re-establishment. Epicormic growth is associated with habitats that promote crown fires rather than fires that only run over the ground surface (He *et al.*, 2012). Some fire-killed species (e.g. *Hakea platysperma*) will occasionally regrow from epicormic buds in response to a low intense fire or some other disturbance (drought, scorched by fire, herbivory).

Epicormic resprouting is a characteristic feature of the eucalypts (Burrows, 2013). The primary function of epicormic regrowth is to rapidly re-establish the crown after fire as the main trunks remain intact, usually because these plants are competing for light with other tree species (Burrows, 2013). However, the many strong stems of lignotuberous species and ecotypes rapidly restore the seed-bearing potential of the plant in an environment where seedling recruitment strongly limits species survival and where light is less limiting for growth (Groom & Lamont, 2011). Epicormic resprouting is a feature of the eucalypts, and in the SouthWest also occurs in trees and tall shrub species within *Banksia* and *Melaleuca*. *B. menziesii* and *B. attenuata* co-occur on the Swan Coastal Plain as epicormic resprouting trees but as a lignotuberous shrub in the more northern Geraldton Sandplains (Groom & Lamont, 2011; He, 2014) as a result of differences in fire characteristics and growing conditions of the two ecosystems.

2.4.4 Rhizomes

Some species resprout from subterranean stems (rhizomes) and root suckers, and thus are not dependent on seedlings to ensure species persistence post-fire. The parasitic tree, *Nuytsia floribunda*, is clonal, producing new shoots from rhizomes that reach up to 100 m around the parent plant (Hopper 2010). A single clump of trees may sometimes occupy several hectares, all ultimately from the one parent plant. *Banksia candolleana*, an outcrossing, creeping, clonal resprouter in the northern sandplains that resprouts from a lignotuber and underground stems, has low seedling recruitment rates and great longevity compared with co-occurring lignotuberous banksias (e.g. *B. attenuata*) (Merwin *et al.*, 2012). Another sandplain banksia, *Banksia elegans*, persists entirely by suckers produced from parent lateral roots (Lamont & Barrett, 1988). This species produces few, if any, fruits and those seeds that are released post-fire rarely become seedlings (Lamont, 1988). *Banksia goodii* is a long-lived (500 years) rare, rhizomotous prostrate species (Fig. 2.6) in the Albany area that may not produce any seeds at all in small populations (Witkowski & Lamont, 1997; Dreschler *et al.*, 1999) – the so-called Allee effect.

This may be due to limited pollinator visits or an increase in the likelihood of selfing (Witkowski & Lamont, 1997).

Rhizomatous and root suckering woody plants are clonal, and may be considered a 'high cost' evolutionary strategy (Fig. 1.2) because a lack of seedling recruitment results in genetically-isolated plants and populations, with restricted evolutionary response capabilities to changing climatic and ecosystem conditions (Fitzpatrick *et al.*, 2008; Merwin *et al.*, 2012). However, clonality enables genetically identical individuals to persist in a habitat for hundreds, sometimes thousands, of years, e.g. 6 400 years for *Eucalyptus phylacis* (Rossetto *et al.*, 1999). This species is only known from a small number of ramets all of which are genetically identical. *Banksia candolleana* is an exceptionally long-lived species (to an estimated 1200 years; Merwin *et al.*, 2012). This means that they can utilize another source of genetic variation (than sexual) involving somatic mutations (time-dependent) and continual bud formation (fire-dependent)—it has been argued that this explains how speciation among resprouters has kept pace with nonsprouters in the SouthWest and other fireprone floras (Lamont & Wiens, 2003).

Rhizomes appear to be a more recent acquisition, in evolutionary terms, in *Banksia* (He *et al.*, 2011), arising from lignotuberous species in response to increased disturbance (increased fire frequency). Clonal species have the capacity to be extremely long-lived, but not always. All *Lepidosperma* species (Cyperaceae) are clonal, resprouting from rhizomes with an estimated age of at least 50 years, possibly 100 years (Barrett, 2013). The lack of any viable seeds or seedlings, characteristic of clonality, suggests an evolutionary trade-off between long-term persistence versus maintaining population genetic diversity, surviving fires by resprouting in an old, climatically stable landscape. Clonal woody species with rhizomes, stem or root suckers may also have a lignotuber and are essentially obligate resprouters.

2.5 Fire-sensitivity

Species that do not survive being burnt are termed fire-sensitive. These species do not resprout post-fire and thus are entirely dependent on recruiting seedlings for their long-term survival (Fig. 2.8). Various names are used to describe these species, the most common being 'nonsprouter', 'obligate seeder' or 'fire-killed'. This is not an adaptation to fire *per se*, so these species must always have other fire-adapted strategies to take advantage of the post-fire environment, including fire-stimulated seed release or germination and/or highly successful seedling recruitment.

Fig. 2.8: Fire-sensitivity. *Hakea polyanthema* (Proteaceae) is a fire-killed (fire-sensitive) species. On the upper left is the dead adult (70 cm wide), on the right is a seedling (10 cm tall) that recruited post-fire. Next to the seedling is an open follicle that protected the two seeds from the fire, releasing them within a few days after the passage of fire onto the loose sand that soon covers it. (*lower left*) Following a fire, all that remains of this *Hakea undulata* (Proteaceae) stand are the dead adult skeletons. Plants average 2 m tall. *H. undulata* relies exclusively on recruiting seedlings post-fire. Seedlings are recruited from seeds germinating during the first winter following a fire, otherwise the seeds perish. (*lower right*) A stand of *Banksia prionotes* (Proteaceae) (3.5 m tall) killed by a surface fire that only scorched them.

2.6 Flammability

Flammability refers to how readily plants burn and can be investigated at scales from individual leaves to the ecosystem. The combined physical and chemical traits of plant species within the landscape, either on the plant or in the litter, will influence whether and how the vegetation will burn, and hence its flammability (Bond & Keeley, 2005). Plant flammability is related to numerous morphological traits, including crown architecture, fine fuel biomass, surface area-to-volume ratio, standing fine litter, stem and foliar moisture levels, periods of climatic dryness and to a variety of chemical properties, including the presence of essential oils (Mutch, 1970; Burrows & McCaw, 1990; White & Zipper, 2010; Pérez-Harguindeguy *et al.*, 2013; Westcott *et al.*, 2014).

The SouthWest flora is highly flammable on all these grounds and is most ignitable during the dry summer-autumn months (Gill, 1975). Myrtaceae, especially *Eucalyptus*, and Rutaceae leaves produce aromatic, essential oils (Fig. 2.9) that quickly ignite and sustain a fire. Eucalypt-dominated woodlands are a feature of the non-sandplain SouthWest landscape. The abundant small leaves (leptophylly) in the shrubland flora (Chapter 9) also have large surface-area-to-volume ratios that increase flammability.

Retention of dead foliage and branches is widespread and is recognised as a flammability trait because it appears to have a primary role in facilitating the spread of fires in the canopy (Midgley, 2013). The SouthWest grasstrees (*Xanthorrhoea* and *Kingia*) retain a 'skirt' of highly flammable dead leaves that combust within three minutes of ignition reaching temperatures up to 1000°C. This ensures that the hidden apical bud is not subjected to periods of prolonged heat, the temperature just above the bud never exceeding a temperature of 70°C (Lamont *et al.*, 2004). The caudex is insulated by the mantle of non-combustible leafbases whose airspaces get filled with molten resin during the fire. The burning leaves in *Xanthorrhoea* release ethylene that promotes conversion of the vegetative bud to a flowering bud by the next spring (Lamont & Downes, 2011). A nutrient-enriched ashbed forms around the plant base but recovery begins months before the winter rains have started to mobilize these nutrients for use by the plant. Two explanations have been proposed for leaf retention by many SWA banksias/dryandras: 1) it ensures that there is sufficient heat to open the follicles for post-fire seed release, and 2) it is an efficient nutrient recycling mechanism for seedlings that establish under the dead parent (He *et al.*, 2011).

The flammability of the SouthWest flora (excluding coastal environments and semi-succulents) has implications for fire intervals, litter accumulation rates, spatial fire patterns, that has led to the evolution and radiation of many plant traits that are related to fire. Combustion of litter and low moisture canopy biomass results in the vaporisation of foliar chemicals and the formation of smoke chemicals that coats the remaining nutrient-enriched ash, as well as releasing a substantial amount of thermal energy (heat pulse) as the fire travels. The heat pulse breaks mechanical dormancy of a number of species with soil-stored seeds (Enright *et al.*, 1997; Tieu *et al.*, 2001) and initiates the release of canopy-stored seeds by killing their seed-protective woody structures. For some *Banksia* species wet-dry cycles are required to initiate fruit opening ensuring that seeds are released with the onset of autumn/winter rains thus protecting their seeds until conditions are optimal for germination (Cowling & Lamont, 1985a).

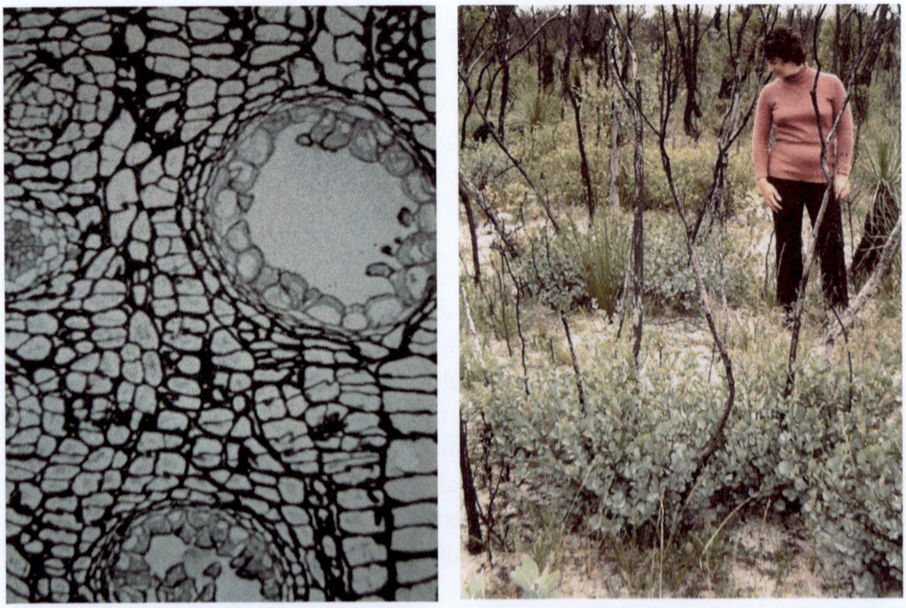

Fig. 2.9: Flammability. (*left*) Longitudinal section through young eucalypt stem showing 100-μm-wide oil glands with the secretory epithelium surrounding the lumen that stores highly flammable essential oils. (*right*) highly flammable mallee eucalypts (*Eucalyptus bupestrium* in foreground) burnt to the ground in the southern Stirling Range and recovering *via* accessory buds stored on lignotuberous root stocks.

2.7 Evolution of Fire-adapted Traits

Table 2.2 summarizes what we currently know about the time of origin of fire-related traits in the SouthWest flora based on molecular phylogenies. The oldest group to show any fire adaptations was the *Petrophile* clade (Proteaceae) at 73.5 Ma, *i.e.* in the Upper Cretaceous, when rainforest covered most of Australia. This demonstrates that fire-prone pockets of vegetation must have existed even at that time. Fire-adapted traits listed include retention of dead flowers and leaves (3 cases as far back as 61 Ma), fire-stimulated flowering (6 cases to 45 Ma), resprouting (15 cases to 69 Ma), soil seed storage (6 cases to 69 Ma), and serotiny (7 cases to 73.5 Ma). Within the Myrtaceae and *Banksia* it is clear that epicormic (tree) resprouting preceded lignotuberous (shrub) resprouting by several million years. While the mallee habit can only be expressed among shrubby, frequently-burnt eucalypts, many tree eucalypts pass through a mallee stage before reaching maturity (e.g. *E. marginata*) while others will coppice from the base if their trunk is cut to the ground, so it is not certain that the mallee growth form is a derived trait. Among banksias it is clear that the ability to recover from fire *via* rhizomes (a trait restricted to banksias in the SouthWest) or root suckers is a derived trait, consistent with declining height of the general vegetation and increasing frequency of fire from the mid-Miocene (as soil provides excellent insulation from fire heat). Nevertheless, rhizomatous Restionaceae can be traced to 61 Ma, indicating that this habit was adaptive even then.

Without exception, the 31 trait-taxa combinations arose ultimately from ancestors in non-fire-prone environments, usually rainforest. Restionaceae is unusual in that its closest families, Centrolepidaceae and Anarthriaceae, are also fire-prone, so one needs to go beyond this order of three families to identify non-fire-prone ancestors. Only eight of the taxa originated at the same time as they became fire-prone (called primary directional selection). The ancestors of the rest were present in fire-prone vegetation for millions of years before they developed fire-adapted traits (called secondary directional selection) or their ancestors were already fire-adapted and they speciated into new genera while retaining those traits (called stabilizing selection), i.e the trait radiated. One wonders how the ancestors survived without the associated fire-adapted trait: the explanation must be that the fire regime changed in some way or some other factor became limiting, necessitating additional adaptive responses for survival. We have already noted the shift from epicormic to lignotuberous to rhizomatous resprouting as the general vegetation declined in stature and fire frequency increased through evolutionary time. Gradually, as leaching continued and seasonality increased, the best opportunities for recruitment would have been confined to the immediate post-fire environment and promoted the implementation of fire-stimulated flowering, soil seed storage and serotiny.

Table 2.2: Time and place of origin of fire-adapted traits in relation to fire-proneness and age of parent clade as determined from trait assignment to molecular time-based phylogenies. Collated by Byron Lamont July-August 2014. Highlighted in blue are genera that developed fire-adapted traits at the same time as they became fire-prone. *includes the relatives Centrolepidaceae and Anarthriaceae, both fire-prone. All ages are for the root of the clade and to nearest 0.5 Ma = million years ago.

Fire-adapted trait	Family	Clade	Time of origin (Ma)	Age fire-prone parent clade (Ma)	Age non-fire-prone parent clade (Ma)	Place of origin	References
Fire-enhancing – dead floret retention – exposed fruits, serotiny – crown fire (fire-stimulated seed release)	Proteaceae	*Banksia*	61	61	61–123.5+	SouthWest	He *et al.*, 2011
Fire-enhancing – dead floret retention – concealed fruits	Proteaceae	*Banksia*	26	61	61–123.5+	SouthWest	He *et al.*, 2011
Fire-enhancing – dead leaf retention	Proteaceae	*Banksia*	26	61	61–123.5+	SouthWest	He *et al.*, 2011
Fire-stimulated flowering	Haemodoraceae	*Blancoa*	18	89.5	89.5+	SouthWest	Hopper *et al.*, 2009; T. He, unpublished
Fire-stimulated flowering	Haemodoraceae	*Haemodorum*	36.5	89.5	89.5+	Australia	Hopper *et al.*, 2009; Lamont & Downes, 2011
Fire-stimulated flowering	Haemodoraceae	*Tribonanthes*	42	89.5	89.5+	SouthWest	Hopper *et al.*, 2009; Lamont unpublished
Fire-stimulated flowering	Haemodoraceae	*Macropidia*	13.5	89.5	89.5+	SouthWest	Hopper *et al.*, 2009; Lamont & Downes, 2011
Fire-stimulated flowering, resprouting – rhizomatous, epicormic	Loranthaceae	*Nuytsia floribunda*	45	45 (possibly 82)	82+	SouthWest	Vidal-Russell & Nickrent, 2008a; Lamont & Downes, 2011

continued **Table 2.2:** Time and place of origin of fire-adapted traits in relation to fire-proneness and age of parent clade as determined from trait assignment to molecular time-based phylogenies. Collated by Byron Lamont July-August 2014. Highlighted in blue are genera that developed fire-adapted traits at the same time as they became fire-prone. *includes the relatives Centrolepidaceae and Anarthriaceae, both fire-prone. All ages are for the root of the clade and to nearest 0.5 Ma = million years ago.

Fire-adapted trait	Family	Clade	Time of origin (Ma)	Age fire-prone parent clade (Ma)	Age non-fire-prone parent clade (Ma)	Place of origin	References
Fire-stimulated flowering	Droseraceae	Tuberous sub-genus	34.5	41	41–50+	Australasia	Yesson & Culham, 2006; Lamont & Downes, 2011
Resprouting – epicormic	Myrtaceae	*Eucalyptus*	57	62	63–90+	Australia	Crisp *et al.*, 2011; M. Crisp, pers. comm.
Resprouting – lignotuberous	Myrtaceae	*Melaleuca*	50	50	50–90+	SouthWest?	Crisp *et al.*, 2011; M. Crisp, pers. comm.
Resprouting – epicormic	Proteaceae	*Banksia*	16	61	61–123.5+	Australia	He *et al.*, 2011; Lamont *et al.*, 2011
Resprouting – lignotuberous	Proteaceae	*Banksia*	13	61	61–123.5+	Australia	He *et al.*, 2011; Lamont *et al.*, 2011
Resprouting – rhizomatous	Proteaceae	*Banksia*	14	61	61–123.5+	SouthWest	He *et al.*, 2011; Lamont *et al.*, 2011
Resprouting – root suckering	Proteaceae	*Banksia elegans*	15.5	61	61–123.5+	SouthWest	He *et al.*, 2011; Lamont *et al.*, 2011
Resprouting – general	Proteaceae	*Hakea*	18	18	35–123.5+	SouthWest	T. He, unpublished; Lamont *et al.*, 2012
Resprouting – lignotuberous, soil seed storage	Proteaceae	*Franklandia*	69	88.5	88.5–123.5+	SouthWest	Sauquet *et al.*, 2009; B. Lamont, unpublished

continued **Table 2.2:** Time and place of origin of fire-adapted traits in relation to fire-proneness and age of parent clade as determined from trait assignment to molecular time-based phylogenies. Collated by Byron Lamont July-August 2014. Highlighted in blue are genera that developed fire-adapted traits at the same time as they became fire-prone. *includes the relatives Centrolepidaceae and Anarthriaceae, both fire-prone. All ages are for the root of the clade and to nearest 0.5 Ma = million years ago.

Fire-adapted trait	**Family**	**Clade**	**Time of origin (Ma)**	**Age fire-prone parent clade (Ma)**	**Age non-fire-prone parent clade (Ma)**	**Place of origin**	**References**
Resprouting – epicormic	Proteaceae	*Xylomelum*	36	45.5	45.5–123.5+	SouthWest	Sauquet *et al.*, 2009; B. Lamont, unpublished
Resprouting – lignotuberous, soil seed storage	Proteaceae	*Stirlingia latifolia*	56.5	88.5	88.5–123.5+	SouthWest	Sauquet *et al.*, 2009; B. Lamont, unpublished
Resprouting – rhizomatous, soil seed storage	Restionaceae	*Lepyrodia* clade	61	101*	101+*	Australia	Briggs *et al.*, 2000; *Bremer, 2002; Litsios *et al.*, 2014, B. Lamont, unpublished
Resprouting – rhizomatous, soil seed storage	Restionaceae	*Apodasmia brownii*	29	101	101+	SouthWest	Litsios *et al.*, 2014
Resprouting – rhizomatous, soil seed storage	Restionaceae	*Hypolaena*	19.5	101	101+	SouthWest	Litsios *et al.*, 2014
Resprouting – rhizomatous, soil seed storage	Restionaceae	*Chordifex laxus*	2	101	101+	SouthWest	Litsios *et al.*, 2014
Serotiny – crown fire (fire-stimulated seed release)	Proteaceae	*Dryandra*	20.5	20.5	61–123.5+	SouthWest	He *et al.*, 2011
Serotiny – crown fire (fire-stimulated seed release)	Proteaceae	*Isopogon*	44.5	44.5	88.5–123.5+	Australia	Sauquet *et al.*, 2009; Lamont & He, 2012

continued **Table 2.2:** Time and place of origin of fire-adapted traits in relation to fire-proneness and age of parent clade as determined from trait assignment to molecular time-based phylogenies. Collated by Byron Lamont July-August 2014. Highlighted in blue are genera that developed fire-adapted traits at the same time as they became fire-prone. *includes the relatives Centrolepidaceae and Anarthriaceae, both fire-prone. All ages are for the root of the clade and to nearest 0.5 Ma = million years ago.

Fire-adapted trait	Family	Clade	Time of origin (Ma)	Age fire-prone parent clade (Ma)	Age non-fire-prone parent clade (Ma)	Place of origin	References
Serotiny – crown fire (fire-stimulated seed release)	Proteaceae	*Petrophile*	49	49	88.5–123.5+	Australia	Sauquet *et al.*, 2009; Lamont & He, 2012
Serotiny – crown fire (fire-stimulated seed release)	Proteaceae	Petrophileae	73.5	73.5	88.5–123.5+	Australia	Sauquet *et al.*, 2009; Lamont & He, 2012
Serotiny – crown fire (fire-stimulated seed release)	Proteaceae	*Lambertia/Xylomelum*	35	45.5	45.5–123.5+	Australia	Sauquet *et al.*, 2009; Lamont & He, 2012
Serotiny – crown fire (fire-stimulated seed release)	Proteaceae	Lambertiinae	45.5	45.5	45.5–123.5+	Australia	Sauquet *et al.*, 2009; Lamont & He, 2012

3 Drought Responses

Southwestern Australia experiences a mediterranean-type climate, with cool wet winters and hot, dry summers. Summer is characterized by months of very little, or no (effective) rain, and hence the plants experience a prolonged drought period. Compounded with air temperatures occasionally rising to 40+°C, the summer-autumn months are the most stressful for plant life, and coping with drought may ultimately determine species survival (Fig. 3.1).

Fig. 3.1: Photographs of two sites during (*left*) and after (*right*) periods of prolonged summer drought. (*upper*) The shrubland vegetation pictured occurs on lateritic soil that is only 30–50 cm deep, and is unable to sustain tree species due to the typical hot, dry summer. A drought stressed tree can be seen at the margin of the shallow soil. Note regreening of foliage from left (autumn) to right (winter). (*lower*) The sand-over-limestone vegetation consists mainly of the fire-pioneer shrub *Acacia pulchella* (Fabaceae) and *Banksia prionotes* (Proteaceae) saplings that grew from soil-stored and canopy-released seeds respectively after a fire in 2008. The *Acacia* plants succumbed to the effects of an extremely prolonged dry and hot summer in 2010 (the brown patches in the understorey, *left*), whereas the *Banksia* saplings (*right*) were able to survive by accessing underground water because they possess a deep root system.

3.1 Herbaceous Drought Avoiders and Resisters

Ephemerals and geophytes are the ultimate drought avoiders in the SouthWest flora. Geophytes are summer dormant, surviving the hot, dry summers as bulbs, tubers, rhizomes and corms protected by the soil (Hnatiuk & Hopkins, 1980; Vaughton & Ramsey, 2001; Parsons & Hopper, 2003; Shane *et al.*, 2009). Even extremely small corms (e.g. the microgeophytic *Stylidium* inhabiting the extremely shallow soils of granite outcrops) are capable of resisting significant desiccation during their summer dormancy (Dixon *et al.*, 1983; Wedge, 2012). *Caladenia* (Orchidaceae) replaces its summer dormant tuber annually, by either forming a replacement tuber, or at the end of a descending structure (the dropper), that places a replacement tuber or daughter tuber at depth (Dixon & Tremblay, 2009). In other geophytes, e.g. *Tribonanthes australis* (Haemodoraceae), a single parent tuber is replaced by a single replacement tuber prior to the next flowering season (Pate & Dixon, 1982).

Lyginia barbata (Anthericaceae) forms dormant roots prior to the onset of the hot, dry summers, regrowing the following winter to a depth of 2-4 m (Shane *et al.*, 2009). The roots of this species are also bound with sand (sand-binding roots or rhizosheaths) (Shane *et al.*, 2011). Sand-binding roots also occur within some species of *Lepidosperma* (Cyperaceae) (Barrett, 2013) and SouthWest Haemodoraceae (Smith *et al.*, 2011). Sand-binding roots are perennial and surrounded by a sheath of sand particles entrapped by root hairs and sometimes mucilage, assisting in water and nutrient uptake in live roots, and minimises root desiccation during summer dormancy (Shane *et al.*, 2011).

A group of herbaceous microchameaephytes, mostly under 10 cm height, may resist and survive the summer heat and drought by raising their shoot system 1-2 centimetres above the soil surface *via* stiff aerial roots (Pate *et al.*, 1984; Pate, 1989). Known as stilt plants, or micro stilt plants (Pate, 1989) they occur in some species of the monocot genera *Stawellia, Hensmania* and *Johnsonia* (Hemerocallidaceae), *Conostylis* (Haemodoraceae), *Borya* (Boryaceae), *Anarthria* (Anarthriaceae) and *Laxmannia* (Asparagaceae) and the dicot genera *Stylidium* (Stylidiaceae) and *Drosera* (Droseraceae).

The carnivorous genus *Drosera*, subgenus *Ergaleium* consists entirely of geophytic stem tuberous sundews (Conran, 2008) from three distinct natural groups (rosette species, fan-leaved species and erect species) that appear to function ecologically as corms. These are summer aestivation organs that also enable the species to survive the summer drought and avoid the heat of a summer fire (Fig. 2.6). Tuberous droseras first appeared 13 Ma and radiated throught the SoutWest during the Miocene (Yesson & Culham, 1996).

3.2 Drought Response of Woody Plants

The southwestern Australian flora has evolved various physiological strategies to cope with drought that are linked to a species rooting pattern, leaf morphology and xylem hydraulics, and are influenced by the level of canopy cover (exposure to the elements), and underlying soil type and groundwater depth. Drought responses can be classified as either avoidance and/or tolerance (Table 3.1) and are related to a species ability to mimimise water loss, maintain cell turgor, and tolerate desiccation. Examples from the Perth sandplain flora are provided in Table 3.2.

Table 3.1: Drought response types characteristic of the SouthWest flora.

Drought Response	Properties
Drought Avoidance	
Water saving	e.g. *Hakea polyanthema* (Proteaceae). This species has needle-like leaves that assist in reducing summer leaf heat loads. In combination with low stomatal conductance values, this reduces leaf water loss. In addition, this species avoids drought by maintaining cell turgor through osmotic adjustment.
Water spending	e.g. *Pericalymma ellipticum*, *Astartea fascicularis* and *Melaleuca preissiana* (Myrtaceae), *Banksia littoralis* (Proteaceae). This response is characteristic of species that occur in wet areas (*i.e.* fringing wetlands) where water is easily accessible all year round. These species typically have high summer transpiration rates, often similar to their winter rates, and thus have high summer water-use requirements.
Drought Tolerance	
Dehydration avoider	e.g. *Banksia attenuata, B. menziesii* and *Adenanthos cygnorum* (Proteaceae). All three species decrease their stomatal conductance before the development of severe drought stress. They also avoid dehydration by maintaining access to water sources *via* their deep root systems.
Dehydration tolerator	*e.g. Hibbertia hypericoides* (Dilleniaceae) is a shallow-rooted species (~ 1 m) that physiologically 'shuts down' to tolerate drought. It does so by severely reducing its stomatal conductance, and photosynthetic ability, and reducing its turgor loss point by osmotic adjustment. The species can also withstand considerable negative water pressures in its xylem (down to –8 MPa).

Sourced from Richards & Lamont (1996); Groom (2002; 2003; 2004); Zencich et al. (2002); Canham et al. (2009); Drake et al. (2013); Bader et al. (2014).

Table 3.2: Physiological case studies from the Perth flora (Groom, 2004).

	***Banksia attenuata* and *B. menziesii* (Proteaceae)** These species co-dominate much of the overstorey within the *Banksia* woodlands that occur on Perth's deep sandy soils, occurring in all topographic locations (dune crests to lower dune slopes) except those areas prone to waterlogging. Both species are dependent on groundwater during the dry summer and possess roots up to 8 m deep. Although both species have access to deep water sources – either groundwater or subsurface soil moisture – they still respond to summer drought by reducing their stomatal conductance and hence leaf transpiration rates. As the soil dries out, the banksias begin to access different sources of soil moisture. In the spring months (September-November) the banksias use subsurface soil moisture; as this dries out during the summer months, they switch to deeper soil moisture sources. *B. menziesii* and *B. attenuata* are trees in the middle and right of the image respectively.
	***Conostephium pendulum* (Ericaceae)** *C. pendulum* is a relatively shallow-rooted species (roots typically less than 1.5 m deep) that relies exclusively on soil moisture rather than groundwater as a source of water regardless of the season. Because of this reliance on limited soil moisture reserves during summer, *C. pendulum* reduces its transpiration rates at the onset of a drought period, thus minimizing its summer water-use requirements. The water relations of shallow-rooted species are strongly influenced by the moisture content of the top 1 m of soil, quickly rehydrating in response to episodic summer rainfall events and break of season (autumn) rains.
	***Adenanthos cygnorum* (Proteaceae)** *A. cygnorum* is a coloniser of soil- and fire-disturbed sites, often occurring in dense stands in the drier banksia woodlands of the Perth sandplain. *A. cygnorum* avoids incurring water stress by maintaining tight control of its summer stomatal conductance, and hence severely reduces its transpiration rates. In addition, *A. cygnorum* possesses a relatively deep root system capable of accessing soil moisture to a depth of at least 7 m.

3.2.1 Diallagy

Summer–autumn foliage colour change (from green to yellow, red, brown or even purple) (Fig. 3.1, 3.2) is most prevalent after a long summer-autumn drought in SouthWest perennial trees and shrubs, especially where they occur on shallow soils. Normal leaf colour within the same leaves returns after 10–15 mm of rainfall. This involves the resynthesis of green chlorophyll pigments that were degraded at the onset of drought, allowing the more heat-tolerant, yellow to red carotene and xanthophyll pigments to dominate leaf colour at this time. This reversible strategy was termed 'diallagy' by George (2002a). George (2002b) recorded 99 SouthWest species in 59 genera and 24 families that display this recoverable drought response. This strategy relies on species being able to dehydrate slowly over increasingly dry months and allowing osmotic adjustment to occur (Richards & Lamont, 1996). The outcome is physiological dormancy, with the plants neither photosynthesising nor transpiring over summer–autumn. *Borya* (Boryaceae) species are small and compact, associated with granite outcrops in the SouthWest (Fig. 3.2). Also known as 'resurrection plants', *Borya* is extremely tolerant of long periods of summer desiccation, rehydrating after the first winter rains (Gaff & Oliver, 2013).

3.3 Classifying Species into Functional Groups

Another way of classifying species is to create 'functional groups'. In physiological terms, a functional group can be defined as a group of species that displays similar properties, adaptations or traits in response to a particular environmental stress or habitat. For SouthWest species, summer drought responses and plant morphologies can explain their habitat preferences and summer survival patterns (Tables 3.1-3.3; Fig. 3.3).

An investigation into characterising the summer drought responses of a SouthWest dry temperate woodland by Mitchell *et al.* (2008a) showed that the responses varying from extremely drought tolerant (summer dormant) to water expensive (Fig. 3.3) were a function of stomatal control. Summer drought-dormant species are characteristically shallow-rooted, small or narrow-leaved, and effectively do not transpire during the heat of the day. Summer dormant species survive only on limited moisture resources obtainable in the top 1 m of the soil profile and respond quickly to increased soil moisture availability caused by episodic summer rainfall events. Four common eucalypt tree species of the SouthWest (*Eucalyptus accedens, E. marginata, E. wandoo* and *Corymbia calophylla*) have been shown to display substantial differences in their leaf water relations (Poot & Veneklass, 2013). Their ability to tolerate summer water stress relates to functional groups based on their rooting patterns (relatively shallow versus deep), soil moisture availability and ability to withstand low minimum leaf and stem water potentials.

Fig. 3.2: In response to prolonged periods of summer-autumn drought, leaves lose their green coloration as chlorophyll degrades, and xanthophyll pigments that are yellow, red or orange predominate. (*upper*) The rare *Banksia ionthocarpa* (Proteaceae) (20 cm tall) near Jerramungup, 150 km NE of Albany, where it occurs on shallow lateritic soil in the green winter-spring condition (*left*) and orange summer-autumn condition (*right*). The same leaves will regreen in response to winter rains. The lack of chlorophyll over summer-autumn is a clue that these species are physiologically dormant at this time of year. (*lower left*) *Banksia tridentata* has a lignotuber with short rhizomes that enable it to recover from dieback due to sudden drought (should it have insufficient time to pass through a colour change period unlike here) or fire. Both species were formerly in the genus *Dryandra*. (*lower right*) *Borya sphaerocephala* (Boryaceae) dominates the shallow sands, and weathered depressions of semi-exposed granite and granite outcrops because it can tolerate prolonged periods of desiccation (dehydration) and summer drought. *Acacia-Allocasuarina* thickets or Wodjil vegetation, dominate the sandy soils surrounding exposed granite of the eastern Avon Wheatbelt where this photo was taken.

In contrast, deep-rooted shrub and tree species can absorb soil moisture or groundwater sources for most, if not all, the year, depending on their position in the landscape (depth to groundwater) and their maximum rooting depth. When summer groundwater levels are not recharged, due to subsequent poor winter rainfall, deep-rooted species are forced to compete for soil moisture sources at depth (Groom *et al.*, 2000). In such situations, it is envisaged that all deep-rooted shrub and tree species would experience considerable water stress.

Table 3.3: Physiological functional groups based on summer drought responses. Information sourced from Mitchell *et al.* (2008a; b).

Functional type	Water use	Drought strategy	Physiological properties	Level of sclerophylly
Tree *Eucalyptus salmonophloia, Eucalyptus capillosa* (Myrtaceae)	All year	Avoidance (deep root system)	Maintains high levels of stomatal conductance during the summer	Low
Shrub—broad-leaved *Grevillea patentiloba* (Proteaceae)	All year	Avoidance (deep root system)	Maintains high levels of stomatal conductance during the summer; Reduced hydraulic transport efficiency (small diameter xylem vessels)	Very high
Shrub—broad-leaved *Dryandra (=Banksia) vestita, Hakea cygna* (Proteaceae)	Summer dormant	Dehydration avoider and tolerators	Increased water-use efficiency, tight stomatal control (low conductance), osmotically adjusts, less negative pressure in the xylem water column	Moderate to high
Shrub—narrow-leaved *Beaufortia micrantha, Melaleuca scabra* (Myrtaceae)	Summer dormant	Dehydration tolerators	Decreased water-use efficiency, osmotically adjusts, more negative pressure in the xylem water column and high wood density	Moderate to very high
Parasitic *Nuytsia floribunda* (Loranthaceae), *Santalum acuminatum* (Santalaceae)	Summer reduced (not dormant)	Obtains water from other plants	Varies between species, *N. floribunda* has a greater reduction in summer stomatal conductance than *S. acuminatum*	Moderate

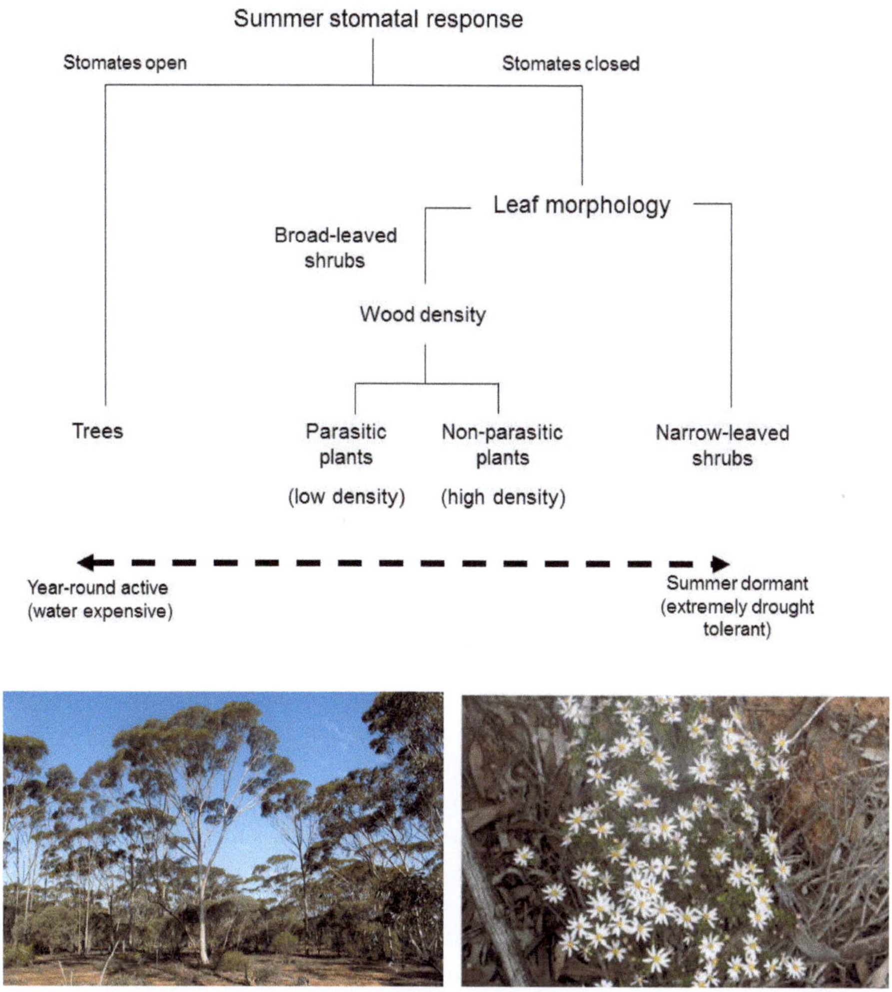

Fig. 3.3: Stomatal responses and leaf morphology are important drivers of a species summer water-use response in temperate SouthWest woodlands. Species response occurs along a water-use spectrum, varying from drought dormant (extremely drought tolerant) to water-expensive. Summer drought responses in the SouthWest flora are a function of position in the landscape, underlying hydrology, species rooting depth and xylem cavitation resistance. (*lower left*) Salmon gum (*Eucalyptus salmonophloia*), a deep-rooted tree species that transpires large amounts of water all year round. Deep roots enable this species to access soil moisture (or groundwater) at considerable depth during dry summers. (*lower right*) *Olearia muelleri* (Asteraceae) a shallow-rooted shrub small species that tolerates drought by becoming summer 'dormant' and displaying significant osmotic adjustment and stomatal closure. Figure adapted from Mitchell *et al.* (2008a).

3.3.1 Functional Significance of Rooting Depth Patterns for Groundwater-Dependent Vegetation

The ability to survive long periods of summer drought is an important determinant of species distribution and is a function of their rooting patterns and physiology, both of which underpin the foundations for a species habitat preference. *Banksia* tree species of the Swan Coastal Plain have the capacity to access groundwater and soil moisture up to 8–9 m depth *via* their deep sinker roots, depending on their position within the landscape. Species classified as obligate phreatophytes (*B. ilicifolia* and *B. littoralis*) are highly dependent on groundwater during the dry summer months. The more drought-tolerant *Banksia* species (*B. attenuata* and *B. menziesii*) are classified as facultative phreatophytes because of their ability to utilize more than one water source, switching to the most abundant type available (Table 3.2).

The proportion of groundwater used varies seasonally, with increased use of groundwater and deeper soil sources occurring with the onset of the dry summer, when rainfall and shallow soil water becomes less abundant (Zencich *et al.*, 2002). Water use patterns of these sandplain woodlands are strongly influenced by the relative contributions of the canopy species (*Banksia* trees and deep-rooted *Adenanthos* shrubs) (Farrington *et al.*, 1989). The amount of water transpired depends on the proportion of canopy cover, amount of winter-spring rainfall received, and has been shown to vary with groundwater depth (Farrington *et al.*, 1989; Veneklass & Poot, 2003). However, on a species basis, tree water use may not change with topography (Zencich *et al.*, 2002).

Shrub species with roots > 2 m deep can tolerate the summer soil drying phase and declining groundwater levels by conserving leaf water while either avoiding (e.g. *Adenanthos cygnorum*) or tolerating (e.g. *Eremaea pauciflora*) the summer drought. These deep-rooted species often have a spreading root system consisting mainly of subsurface laterals similar to that of *Banksia* (Dodd *et al.*, 1984) surviving periods of summer drought by accessing soil moisture resources at greater depths. Deep-rooted shrub species are therefore capable of tolerating short-term declines in soil moisture and groundwater levels, switching from groundwater to soil moisture reserves when groundwater, and the associated capillary fringe, becomes unavailable (Groom, 2003; 2004). In contrast, shallow-rooted shrub species are able to tolerate severe summer soil water deficits. This allows these species to survive the dry summer without accessing groundwater or soil water sources greater than 1 m depth. The exceptions are species that are restricted to lowlying depressions or seasonally waterlogged areas that are reliant on subsurface soil moisture or groundwater to sustain their high summer water-use (Groom, 2003).

Abstraction of groundwater from shallow, unconfined aquifers underlying the *Banksia* woodlands continues to impact on sandplain hydrology, resulting in shifts in species distributions along topographic gradients (Groom *et al.*, 2000; 2001). Declining groundwater levels caused by abstraction, in association with poor winter rainfall-induced groundwater recharge, have the potential to change the vegetation structure and composition of these sandplain woodland, particularly in interdunal, low-lying areas that form seasonally wet damplands (Groom *et al.*, 2000; Sommer & Froend, 2011; Barron *et al.*, 2014). Species responses to soil drying and declining groundwater levels form the basis for understanding how Perth's *Banksia* woodland species will tolerate and survive long-term hydrological changes (Sommer & Froend, 2014), and this can be explained by functional differences in rooting patterns.

3.3.2 Woody Perennial Rooting Patterns

Strategies of the trees and shrubs to survive prolonged periods of dryness are dependent on the depth and spread of their root systems. Dodd *et al.* (1984) classified five functional types occurring within the SouthWest sandplain shrub flora, depending on their ability to access soil moisture or the groundwater capillary fringe at depth. Most shrubs, and all tree species, possess dimorphic root systems, consisting of spreading laterals, a central tap root and multiple vertically-aligned sinker roots (Lamont & Bergl, 1991; Pate & Dixon, 1996). For fire-pioneer shrub, *Acacia pulchella* (Fabaceae), these sinker roots are relatively shallow (< 1 m) and subsurface laterals spread no more than 1 m (Dodd *et al.*, 1984). In contrast the deeper-rooted shrub, *Jacksonia furcellata* (Fabaceae), has laterals that spread to 5 m with a depth > 2 m (Dodd *et al.*, 1984). Some shrub species only have shallow spreading laterals (depth and spread < 50 cm), e.g. *Hemiandra pungens* (Lamiaceae), others only have a deep sinker (or tap) root, e.g. *Petrophile linearis* (Proteaceae), whereas others have a specialized root system that is characterized by shallow branched roots (*e.g. Conostephium pendulum* (Ericaceae) with no sinker or lateral roots.

High levels of floral biodiversity exist regionally and locally within the SouthWest because of the various rooting pattern strategies employed among co-occurring species. Where summer moisture is always limited (e.g. shallow soils, and highly permeable sandy soils) there is extensive competition to meet species' summer water-use demands, to the extent that no one species can outcompete others. Shallow-rooted species are often extremely drought tolerant, or summer dormant, particularly in the harsh sandplain environments where summer soil moisture contents are often much less than 1%. These species tolerate prolonged periods of summer drought through restricting water use, e.g. closing their stomates, osmotically adjusting, and are capable of withstanding extremely low water pressures and increasingly frequent cavitation events within their xylem, and quickly rehydrate in response to episodic summer rainfall (Zencich *et al.*, 2002). Shallow-rooted species in the annually wetter forests are potentially vulnerable to increasingly prolonged droughts because of the difficulties in accessing moisture from the highly developed, layered lateritic soils (Crombie *et al.*, 1988; Crombie, 1992).

Rooting patterns of deep-rooted trees and shrubs enables different sources of moisture to be utilized as water sources become limiting (Canadell *et al.*, 1996). Roots of different species often intermingle and co-occurring species from the same genus (cogeneric) may display similar seasonal water-use patterns (Lamont & Bergl, 1991) despite differences in fire life-history traits. During the SouthWest wetter winter months, the dimorphic-rooted *Banksia prionotes* (Proteaceae) utilizes water uptaken by subsurface roots that feed into the large laterals, whereas in the drier summer months water is obtained from groundwater and their associated capillary fringe (Dawson & Pate, 1996). *B. prionotes* also is capable of hydraulic lift, where soil water available at depth is transported to the drier subsurface *via* its root system (Burgess *et al.*, 2000). The redistribution of water to the subsurface soil rehydrates the fine water and nutrient absorbing roots (Horton & Hart, 1998; Burgess *et al.*, 1998). There is the potential for some deep-rooted species to demonstrate hydraulic redistribution following a rainfall event whereby subsurface water is moved to deeper layers of the soil profile (Burgess *et al.*, 2001) where it is not prone to evaporation. Deep-rooted species will maintain access to the seasonally shifting groundwater capillary fringe all year round, where the underlying superficial aquifer is less than 10 m deep (Canham *et al.*, 2012), except when summer water levels drop faster than roots can grow.

3.4 Stomatal Location—a Drought Response?

Stomates may occur on the underside of the leaf (bifacial or dorsiventral), or on both sides (isobilateral), or isolateral if terete. Stomates can be sunken (see Fig. 3.5), or occur in crypts, grooves or surface pits (Jordan *et al.*, 2008) that provide additional protection from a drying and turbulent atmosphere. This has been tradtionally viewed as a means of limiting leaf water loss *via* transpiration by providing a layer of relatively still air to water vapour movement. Sunken stomates, and other means of stomatal protection, within the Proteaceae arose at least 11 times in arid environments through the family's evolution, but not always with dry climates (Jordan *et al.*, 2008). Thus their primary function may not always be to reduce transpiration (Roth-Nebelsick *et al.*, 2009). Studies on the stomatal crypts of *Banksia* suggest that crypts may also serve to facilitate the diffusion of carbon dioxide into thick leaves (Hassiotou *et al.*, 2009). Nevertheless, the thickest and densest leaves are in the driest regions of the SouthWest (Lamont *et al.*, 2003) so that encrypted stomates are also associated with the thickest and most drought-tolerant leaves. It is worth noting that the crypts close as the leaf dries and shrinks during drought so preventing further water loss since the stomates are confined to these grooves in *Hakea invaginata* and *Borya* leaves and in *Casuarina* branchlets (Fig. 3.4). Thus, encrypted stomates are associated with thick leaves and both are best interpreted as efficient responses to drought, first appearing in *Allocasuarina*/*Casuarina* 39 Ma (Crisp *et al.*, 2004).

Fig. 3.4: (*upper left*) Flowering branch, 25 cm tall, of *Hakea invaginata* showing its terete leaves. (*upper right*) Part cross section of *H. invaginata* leaf, 2 mm diameter, showing stomates restricted to the grooves with a cuticle 2–3 times thicker than the epidermis and tannin-bearing, wall-thickened supporting parenchyma, all indicating anatomical adaptations to drought. The palisade is stained green and the sclerenchyma red. (*middle right*) Line drawing of cross section of *Borya sphaerocephala* showing four grooves (g) that penetrate the chlorophyll-bearing tissue (broken lines) with an otherwise continuous band of sclenchyma beneath the epidermis (e) (from Gaff *et al.*, 1976). (*lower*) Scanning electron micrograph of branchlet (cladode), 2 mm diameter, of *Casuarina glauca* showing grooves, which run up to the scale leaves, true leaves are absent), to which the stomates and photosynthetic tissues are restricted and filled with persistent hairs that also serve to decrease the rate of water loss. The grooves close as the branchlet dehydrates preventing further water loss.

The remarkably diverse genus *Acacia* (Fabaceae: Mimosoideae) in the temperate and semi-arid SouthWest has surface stomates on the lower (shaded) side of bipinnate compound leaves, or on both sides (amphistomatic) of phyllodes (vertically-aligned, leaf-like petioles). Groso *et al.* (1994) recognised 6 stomatal types within the genus that differ in the positioning of the surrounding subsidiary cells.

3.5 Case Study: Ecophysiology of Dimorphic Leaves in *Hakea trifurcata*

Hakea trifurcata (Proteaceae) is heterophyllous, meaning it possesses more than one leaf type within the same individual—in this case it is dimorphic with broad/incurved and terete/needle-like leaves. Both leaf types are usually present on the same branchlet (Fig. 3.5). Typically 1–2 broad leaves form first during the annual growth period (late winter) followed by many terete leaves (spring). Broad leaves photosynthesise and transpire more water per unit mass than terete leaves and experience greater tensions within their xylem water column (Groom *et al.*, 1994a).

When rates of transpirational water loss over an autumn day (April) are compared among broad and needle (terete) leaves on the same branches of *H. trifurcata*, the broad leaves lose four times as much water as the terete leaves for the same leaf weight. Photsynthetic rates show similar differences. Transpiration (and photosynthesis) peaks in the early morning and mid-late afternoon and it approaches zero in both leaf types during the hottest time of day (the socalled 'midday depression'). The size and shape of the terete leaves reduce their heat load and enable the leaves to sustain lower summer leaf temperatures and better summer water-use efficiency, characteristics commonly associated with drought-tolerant leaves. Broad leaves are characteristic of leaves that keep their stomates open during periods of water and heat stress. As the first formed at the start of spring they serve to maximise carbon gain for subsequent growth.

Fig. 3.5: (*left*) *Hakea trifurcata* (Proteaceae) in flower showing the two leaf types that it produces on the same branch. The terete leaves can be either simple or 2–3-pronged. Broad leaves range from 1.5 to 5 cm in length. (*upper right*) *Hakea erinacea* is in the same taxonomic group as *H. trifurcata* but only possesses 2-cm-long, pronged needle leaves. (*lower right*) Transverse section through a semi-terete leaf of *H. erinacea*. Arrows indicate the location of sunken stomates positioned below the epidermis (outer layer of cells) in crypts all around the leaf (amphistomatous). Section is 2 mm wide.

4 Carnivorous Plants

Carnivorous plants produce elaborate and highly specialized traps to capture invertebrate prey. These traps arise from modified leaf structures. However, to be truly carnivorous, a species must also demonstrate an ability to produce digestive enzymes (that can break down invertebrate tissues) and absorb the resulting nutrient-enriched solution. Some genera demonstrate protocarnivory (they have some, but not all features of carnivory). For example glands on the peduncle and flowers of some *Stylidium* species produce digestive enzymes (Darnowski *et al.*, 2006), but experiments are still required to determine if nutrient uptake occurs. Resin glands on the scapes, bracts and leaves of *Grevillea leucopteris* capture many insects but there is no mechanism for their digestion.

Western Australia's carnivorous plants are confined to five genera (*Drosera, Byblis, Cephalotus, Utricularia, and Aldrovandra*) that each displays a different capturing technique, with the greatest number of species in Australia occurring in the SouthWest. Carnivorous plants typically grow in habitats that are nutrient-impoverished ranging from highly leached soils to bogs and swamps where organic compounds acidify the water, reducing the rate of decomposition and hence nutrient availability. *Aldrovandra* is a monotypic genus and has only recently been recorded in southwestern Australia near Esperance, but is not in great abundance. There is an additional record of the genus occurring in the Kimberley.

4.1 *Drosera*

Drosera (Droseraceae) has its worldwide centre of diversity in the SouthWest and exhibits a range of growth forms (Rivadavia *et al.*, 2003; Yesson & Culham, 2006; Lowrie, 2014). SouthWest species are mostly perennial and are classified into three subgenera: the tuberous *Eryaleium* includes species that have rosettes of leaves (section *Erythrorhizae*), leaf clusters on short stems (section *Stolonifera*) or climb (section *Ergaleium*), and the subgenus *Bryastrum* (pygmy droseras) (Fig. 4.1). The annual, sometimes ephemeral, *D. glanduligera*, is the sole species of subgenus *Coelophylla* and possesses a combination of sticky and snap trap tentacles, unlike other species. Snap traps occur in the venus fly trap [*Dionaea muscipula* (Droseraceae) native to North and South Carolina, USA] but the mechanism here is quite different. Originating 40 Ma (Yesson & Culham, 2006), *D. glanduligera* inhabits moist low-lying, well drained soils, and is a relict of these ancient mesic Eocene landscapes.

The leaves bear sticky, multicellular hairs (tentacles) that capture arthropods, attracted by colours, the glistening mucilage produced by the tentacles, or scent (Gibson & Waller, 2009). These tentacles respond to the smallest touch, causing the tentacles to slowly bend towards the prey and downwards. This response is transmitted to the other tentacles. The greater the prey struggles, the more widespread is tentacle movement towards it. The captured prey becomes covered in mucilaginous fluid secreted by the glands that have captured it. *Drosera* is derived from the Greek *drosos* (dewy) and refers to the glandular hairs that give the appearance of being covered in dew, hence the common name, sundews

Stickiness of the tentacles (Fig. 4.2) is due to the production of mucilage that contains proteolytic and other hydrolysing enzymes that break down proteins and other compounds in the prey. Chitinous exozymes (chitinases) are also produced by bacteria that break down the prey's tough exoskeleton (Fig. 4.3). Absorption of digested products occurs by the same gland that secretes the digestive enzymes, accounting for 50-80% of drosera nitrogen uptake (Dixon, Pate & Bailey, 1980) and utilization (Schulze *et al.*, 1991). After a few days the tentacles resume their original position, the undigested remnants drop away and the modified leaf is ready to capture more prey.

Fig. 4.1: *Drosera* species display great diversity in their habit. (upper *left*) *D. rupicola* in flower showing peltate leaves with glands in the leaf concavity (5 mm wide). (*upper middle*) the SouthWest is a biodiversity centre for pygmy droseras. Pictured is *D. nitidula* (1 cm diameter) displaying relatively long glandular tentacles arising from the modified leaves. (*upper right*). Close up of the 'dewy' stalked glands (tentacles) of *D. auriculata* with remains of a digested insect in the lamina (lamina 5 mm wide). (*lower*) erect (*D. porrecta*), climbing (*D. erythrogyne*) and rosette (*D. rosulata*) droseras. Many species produce new shoots from an underground tuber in response to the winter rains. Images of *D. auriculata, D. erythrogyne, D. nitidula* and *D. porrecta* provided by Allen Lowrie.

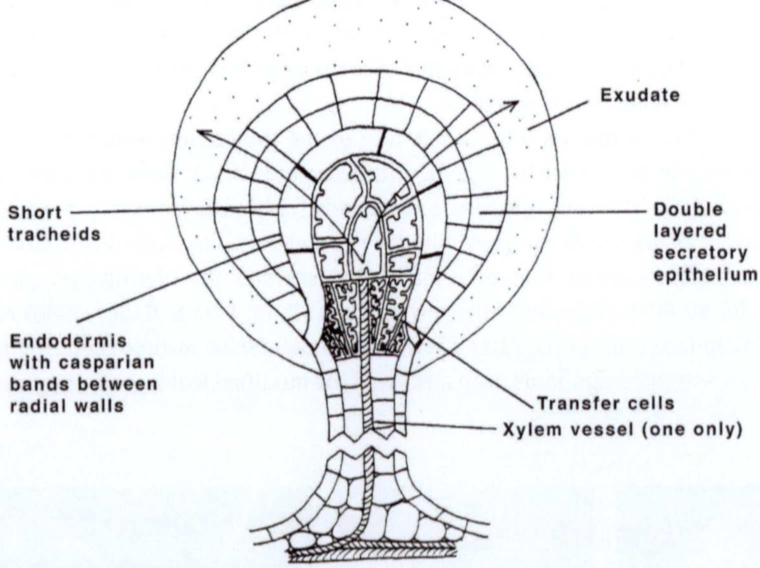

Fig. 4.2: Longitudinal section through a *Drosera* tentacle. The glandular head of a tentacle is usually reddish and covered with droplets of mucilage. The glandular head consists of external and internal layers of secretory cells and tracheids in the centre. The tracheids are separated from glandular cells by a row of barrier cells consisting of cutinized cell walls that appear to control transport. In the basal part of the tentacle there are some outer (epidermal) and inner (parenchyma) stalk cells and a single vessel connecting the glandular head with the stalk and hence the leaf. Head 0.5 mm wide.

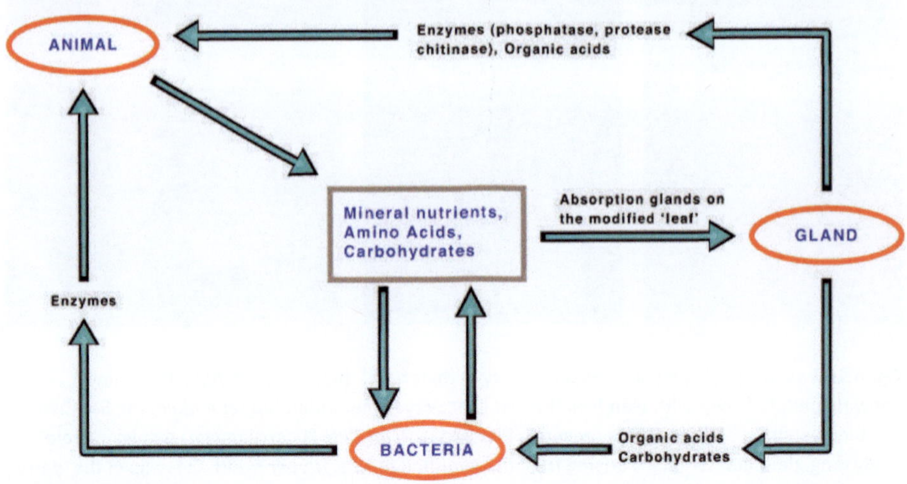

Fig. 4.3: Schematic representation of the digestive and absorption processes involving exudates produced by a *Drosera* tentacle. *Drosera erythrorhiza* can capture on average 80 arthropods cm^{-2} day^{-1} in mid-spring (Dixon *et al.*, 1980), and are a more effective source of N than soil-supplied N.

4.2 *Byblis*

Two species of *Byblis* (Byblidaceae) inhabit southwestern Australia (*B. gigantea, B. lamellata*) (Lowrie *et al.*, 2002). Both are perennial subshrubs, regenerating each year from the stump of the previous year's stem. The remaining four species occur in northern Western Australia and are all annual herbs, relying on seed production for species survival. *Byblis* is named after a princess of classical Greek mythology who fell in love with her brother; she eventually hanged herself and turned into a fountain. This refers to the stem and leaves of the plant that are covered with glandular hairs glistening in the sun 'like dripping tears'.

The leaves and stems of all *Byblis* species are densely covered with glandular hairs that secrete a mucilaginous fluid (Fig. 4.4). These serve to attract small insects that are ensnared when touching the sticky secretions. Unless they are strong enough to escape, the insect prey either dies from exhaustion or asphyxiates as the mucilage envelops it. Unlike the sundews, however, *Byblis* are unable to bend their tentacles or leaf lamina to trap or digest prey. As a result, they are classified as a "passive trap".

Byblis has two types of glands - stalked and sessile glands (Fig. 4.4). Stalked glands capture and twist down after prey capture. Sessile glands secrete the digestive juices and absorb the products of digestion, and are five to ten times as numerous as the stalked glands.

Fig. 4.4: (*left*) *Byblis lamellata* growing in seasonally damp sands at Eneabba. Glands cover the needle-shaped leaves, stems and sepals. Above plant is 25 cm wide. (*right*) Electron micrograph of a *Byblis* leaf showing the stalked and sessile glands covering the surface. Stomates (leaf pores) on the leaf surface are also visible. Leaf is 2 mm wide.

4.3 *Cephalotus*

Cephalotus has only one species and that is the only species in the family Cephalotaceae (thus it is a monotypic family). Locally known as Albany pitcher plant, the species grows in peaty swamps along the south coast of southwestern Australia. The name *Cephalotus* is derived from the Greek *kephale* (having a head). This refers to the head of pollen at the end of the stamen.

The trap, a modified leaf, lures prey by a combination of coloured markings on the rim and lid, its odour (decaying prey) (Fig. 4.5), presence of a reward (nectar along the rim) and transmitted light through the translucent lid and wall tissues that suggest a passage (Parkes & Hallam, 1984). It is an effective arthropod-trapping device because:

1. The reddish-purple mouth has a slippery surface ribbed with downwardly-pointing teeth arching over the pitcher, so that an insect tends to slip into the pitcher rather than escape from it.

Fig. 4.5: *Cephalotus follicularis.* (*upper left*) Note the mature (*left*), immature (*middle*) and embryonic (*right*) pitchers. These pitchers are modified leaves. Mature pitcher is 4 cm long. (*upper right*) Drawing of a longitudinal section through a *Cephalotus* pitcher. Pitcher is 4 cm long. (*lower left*) insect-pollinated flowers. The species epithet refers to its hooded stamens. (*lower right*) *Cephalotus* plants occur near or in swamps and bogs in the south coastal SouthWest. Plant pictured is 15 cm wide. Images provided by Melissa Darling (*upper left*), Allen Lowrie (*lower left*) and Aaron Gove (*lower right*).

2. The inside rim (collar) curves downward to form a ledge and is glandular and slippery making it difficult for an insect inside the pitcher to climb out.
3. Smooth pitcher walls.
4. The translucent lid and walls make it difficult to distinguish the exit for any insect flying into the pitcher.
5. The liquid pool at the bottom of the pitcher also reflects light, as well as capturing and drowning its prey.

Arthropods that fall into the pitcher drown in the digestive fluid. Glands in two mounds on the pitcher wall exude digestive enzymes and absorb the resulting nutrient-enriched solution. Large *Cephalotus* plants obtain, on average, 26% of their nitrogen requirement from captured prey (Schulze *et al.*, 1997).

4.4 *Utricularia*

Of the 40 species known in Western Australia, thirteen occur in the SouthWest. Although *Utricularia* species mostly occur in moist peat, some occur on sand and in depressions on rock outcrops, while others are completely aquatic (with only the flowers appearing above water). They can survive almost anywhere where there is fresh water for at least part of the year. Many Australian species will grow only during the wet season, reducing themselves to tubers only 10 mm long and remaining dormant during the dry season.

Utricularia possesses the most complex traps of all carnivorous plants (Fig. 4.6). Called 'bladders' (utricles) they are matchhead in size, and are modified leaf segments (aquatic species) or whole leaves arising directly beneath a rosette of normal leaves (non-aquatic species). The name *Utricularia* is derived from the Latin *utriculus* (small skin or leather bottle) in reference to the bladder. Each trap is partially water-filled, and has a door surrounded by sensitive hairs that direct prey to the opening. The trapdoor opens inwards upon irritation by a passing animal such as a mosquito larva. After the prey brushes against the trigger hairs, it is sucked in because of the release of negative pressure maintained inside the utricle, engorging several times its resting volume. Traps are connected to the plant *via* stolons, stalks or leaves. Plants may even be rootless.

After firing, the trap restores the negative pressure by removing water from the lumen until the original compressed shape is returned. After this process is completed, which lasts about 30 minutes, the trap is ready to fire again.

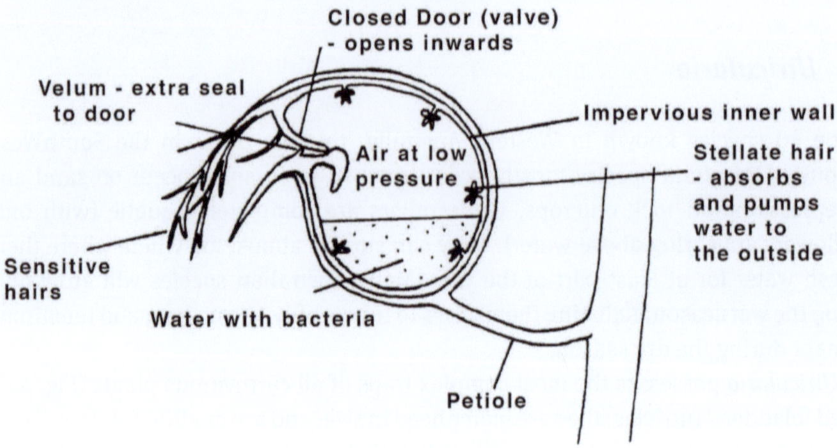

Fig. 4.6: (*upper*) *Utricularia menziesii*, 3 cm tall. (*lower*) Diagram of a longitudinal section through a *Utricularia* bladder. The trap is a hollow utricle, mostly two cells thick, partly filled with water, and under negative pressure. Traps are usually 1–4 mm long.

5 Parasitic Plants

Parasitic plants form a close connection with the vascular system of their host plant(s) through a specialised structure known as a haustorium (plural haustoria), that physically connects the parasite to the host, providing a vascular conduit for water and nutrients (and sometimes sugars and amino acids) from host to parasite. They are dependent on their host for their supply of mineral/inorganic nutrients, water and/or organic compounds, although the degree of host dependency varies greatly between species. Almost 1% of the world's flowering plants are parasitic (~3,000 species), and it is estimated that parasitism has appeared at least 11 times during the evolution of angiosperms (Barkman *et al.*, 2007).

A hemiparasite [*half parasite*] obtains water and mineral nutrients from the host xylem but is also photosynthetic. Some hemiparasites also obtain limited amounts of carbohydrates from the host phloem. The haustoria are either attached to the host's shoot or root. Holoparasite [*whole parasite*] is where the parasite is non-photosynthetic and obtains water and nutrients from the host xylem and carbohydrates from the host phloem. The parasite occurs within the host's shoot or root tissues. Holoparasites are less common than hemiparasites in the SouthWest flora (Table 5.1).

Table 5.1: Parasitic genera in southwestern Australia. Root = root parasite, shoot = shoot parasite. Data obtained from *Florabase* (florabase.dpaw.wa.gov.au), accessed October 2013.

Loranthaceae (hemiparasites)	**Lauraceae** (hemiparasites)
Amyema (12 species; shoot)	*Cassytha* (9 species; shoot)
Lysiana (2 species; shoot)	
Nuytsia (1 species; root)	**Olacaceae** (hemiparasites)
	Olax (4 species; root)
Santalaceae (hemiparasites)	
Santalum (4 species; root)	**Apodanthaceae** (holoparasites)
Anthobolus (1 species; root)	*Pilostyles* (3 species; shoot)
Choretrum (3 species; root)	
Exocarpos (3 species; root)	**Orobanchaceae** (holoparasites)
Korthalsella (1 species; shoot)	*Orobanche* (1 species; root) (exotic)
Leptomeria (12 species; root)	**Convolvulaceae** (holoparasites)
Spirogardnera (1 species; root)	*Cuscuta* (1 native species, 3 exotic; shoot)

5.1 Stem Hemiparasites - Mistletoes

Worldwide, only three of the 75 genera in the family Loranthaceae do not display the typical 'mistletoe' habit, *i.e.* clumps of foliage attached to host branches (Wilson & Calvin, 2006). One of these is the WA christmas tree (*Nuytsia floribunda*). Two aerial mistletoe genera occur within the SouthWest flora–*Amyema* and *Lysiana* (Fig. 5.1), of which 80% of the species are endemic. *Korthalsella arthroclada* (Santalaceae) also displays the mistletoe habit, and only parasitises *Melaleuca lanceolata* beside salt lakes near Eneabba.

Mistletoes have no root system, and germinate without the need for free water. They parasitize trees and tall shrubs and may be host specific. The seeds are surrounded by a sticky mucilage that enables them to adhere to a host branch. If deposited on a dead branch, the mistletoe is free-living for five months before it starts to die, and unable to survive on non-host living branches for more than 6 months (Yan, 1993b). Although mistletoes possess green foliage and thus produce their own carbohydrates, they are entirely dependent on their host for their water and inorganic nutrient supply. Mistletoes have higher transpiration rates than their host plants. Higher transpiration rates are essential to 'pull' water from the host plant. They also possess higher concentrations of inorganic nutrients, especially potassium and (uniquely in the SouthWest) sodium that also increases the osmotic concentration on the mistletoe side of the union.

5.1.1 Host Mimicry

Because mistletoes have high levels of nutrients (especially potassium, sodium, nitrogen, phosphorus) in their leaves (Lamont, 1983), they are relatively nutritious (compared with their hosts) and are an easy target for herbivores. Many mistletoes have developed leaves that mimic the foliage of the host plant (Fig. 5.1). Some are restricted to a few species or predominately parasitize a particular genus (Yan, 1990) (e.g. *Amyema preissii* on *Acacia; Amyema melaleucae* on *Melaleuca*), others have many hosts (e.g. *Amyema miquelii* has 125 host species). Host specificity is a function of bark and xylem resistance to haustorial penetration (Yan, 1993c).

There are many tree and tall-shrub species that are not susceptible to mistletoe parasitism (see Table 5.2 for a list of mistletoe hosts). Of the ~25,000 vascular plants in Australia 870 are known to serve as host species (Downey, 1998).

Amyema is a xylem-tapping genus (Fig. 5.1) with no known ability to abstract phloem-borne solutes from its hosts. Mistletoes establish xylem-xylem contact early in the development of the endophyte, although direct xylem contacts are relatively few due to the large numbers of parenchymatous (non-specialised) cells at the interface. They derive no organic nutritional benefit apart from small amounts of carbon compounds in the transpiration stream.

Fig. 5.1: (*upper left*) *Amyema benthamii* (*centre*) hosting on *Brachychiton gregorii* (Malvaceae). (*upper right*) *Amyema gibberula* (centre) *on Hakea recurva* (Proteaceae). Note that both species show host mimicry in their foliage. (*middle*) Flower (2.5 cm long) and fruit (8 mm long) of the mistletoe *Lysiana casuarinae* (Loranthaceae). As the flower opens the petals fold back to reveal the long green stamens. The fruit (a berry) ripens to a red or black colour. *L. casuarinae* has long (up to 9 cm) needle-like leaves, and in this case was hosting on *Acacia acuminata* (Fabaceae). (*lower left*) Drawing of a cross section through a young *Amyema* haustorium and the infected stem of a host. Host stem diameter is 3 cm. (*lower right*). Two *Amyema* seeds (6 mm long) deposited simultaneously by a mistletoe bird, though only one is attached to the host branch. The green haustorium is visible on each seed, growing towards the branch (negative geotropism).

Table 5.2: List of southwestern Australian *Amyema* and *Lysiana* species. Based on information obtained from the WA Herbarium online database, *Florabase*, accessed October 2013. Examples of SouthWest hosts and number of host species obtained from Downey (1998).

Amyema benthamii (31 host species)
(on a variety of hosts; mostly Eremaean)
- *Brachychiton gregorii* (Malvaceae)

Amyema gibberula (20 host species)
- *Acacia acuminata*
- *Hakea recurva*

Amyema linophylla (8 host species)
(on several species of Casuarinaceae)
- *Acacia aneura*
- *Casuarina obesa*

Amyema melaleucae (8 host species)
(exclusively on *Melaleuca*)
- *Melaleuca lanceolata*
- *Melaleuca laxiflora*
- *Melaleuca pauperiflora*

Amyema miquelii (125 host species)
(mainly on *Eucalyptus* and some *Acacia*)
- *Corymbia calophylla*
- *Eucalyptus gomphocephala*
- *Eucalyptus loxophleba*
- *Eucalyptus salmonophloia*
- *Eucalyptus wandoo*
- *Santalum spicatum* (also parasitic)
- *Melaleuca lanceolata*

Amyema miraculosa (41 host species)
- *Acacia aneura*
- *Santalum acuminatum* (also parasitic)

Amyema preissii (73 host species)
(mainly on *Acacia*)
- *Acacia acuminata*
- *Acacia cyclops*
- *Acacia hakeoides*
- *Acacia hemiteles*
- *Acacia saligna*

Lysiana casuarinae (17 host species)
(on a variety of hosts including
(*Allo*)*Casuarina*, *Acacia* and *Exocarpos*)
- *Allocasuarina huegeliana*
- *Casuarina obesa*
- *Acacia acuminata*
- *Acacia aneura*
- *Exocarpos aphyllus* (also parasitic)

5.1.2 Seed Dispersal

Amyema and *Lysiana* fruits are a succulent berry (Fig. 5.1) that is attractive to birds, including the Australian mistletoe bird (*Dicaeum hirundinaceum*: Dicaeidae) (Yan, 1993a). The conspicuous berry is swallowed whole (minus the outer rind) and many berries can be consumed in one feeding session.

The mistletoe bird defecates the sticky seeds onto a host branch (Fig. 5.1). Because the seeds are surrounded by a sticky mucilage layer, the bird has developed a special 'bottom-wiggle' or 'zig-zag' dance to wipe off the voided seed clinging to its cloaca, which ensures that seeds are deposited onto a branch. The mistletoe bird has an extremely short intestinal system, and seeds are defecated 30 minutes after being consumed. This explains why mistletoes tend to be crowded on and around a given host plant.

Once *Amyema* seeds have matured, they remain dormant in the berry (Fig. 5.2). Excision from the berry is the main trigger to promote germination and growth, which may commence within 10 hours of the excision event occurring (*i.e.* removal by the mistletoe bird) (Lamont, 1983). The high carbon dioxide (CO_2) concentration that occurs within the berry is sufficient to maintain dormancy while the seed remains within the berry (Lamont, 1982).

The outer rind of the berry is impervious to gas movement in or out of the berry, but allows 40% of photosynthetically active radiation (light at 400–700 nm wavelengths) to pass through that is used by the chlorophyll-bearing embryo to undergo limited photosynthesis during the day (Lamont & Perry, 1977). However, respiration occurs both night and day so CO_2 production exceeds oxygen (O_2) production overall, preventing further growth of the embryo.

This causes a buildup of CO_2 in the berry at concentrations greater than that in the earth's atmosphere. When the seed is excised (*i.e. via* the dispersal process involving the mistletoe bird), it becomes exposed to low CO_2 and high O_2 concentrations and can germinate. Light is also important in assisting *Amyema* germination, as the sugars produced *via* photosynthesis are required to support tissue growth (Lamont & Perry, 1977). It must rely on its own store of water and nutrients until the young haustorium penetrates the host bark and reaches its xylem.

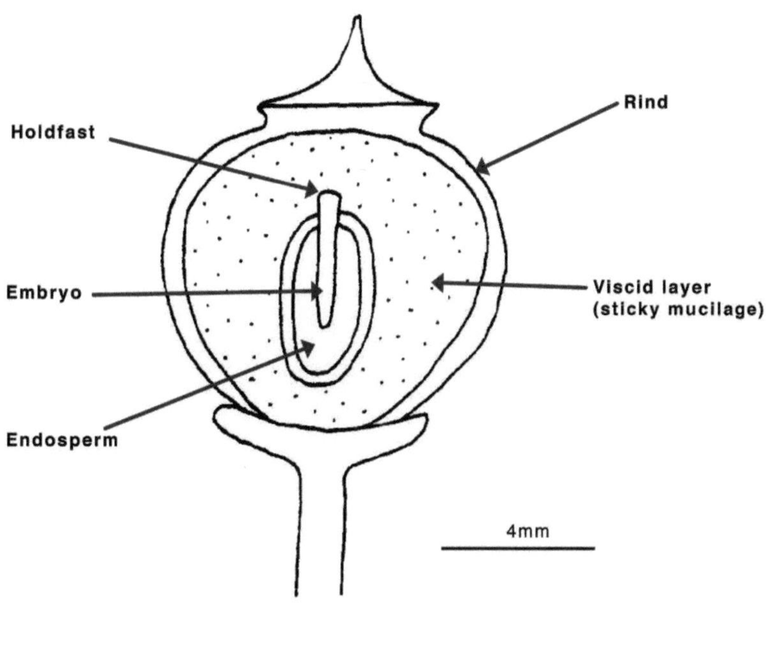

Fig. 5.2: Diagram of a longitudinal section through a mature *Amyema* berry.

5.2 Root Hemiparasites

5.2.1 *Nuytsia Floribunda* (Loranthaceae)

Nuytsia floribunda is the only species in its genus and its biology is so distinctive it was once considered to be a monotypic family (Hopper 2010). The species epithet, *floribunda*, meaning 'lots of flowers', refers to the spectacular display of orange flowers that occurs annually between November to January (Fig. 2.4), and hence is known locally as the Western Australian christmas tree. *Nuytsia* is a tree mistletoe that parasitizes a range of plant species (Calladine & Pate, 2000), and is ancestral within the Loranthaceae: it originated 45 Ma and is a true relict of the Eocene Epoch (Wilson & Calvin, 2006; Vidal-Russell & Nickrent, 2008a; b).

The fruits are three-winged and wind dispersed (Fig. 5.3) that is unique within the order Santalales. The papery wings also have an important function in absorbing and channelling rainwater to the germinating seed. However, regeneration is mostly in the form of stem suckers arising from long underground rhizomes. These stem suckers develop a shallow root system that ultimately bears the fine haustoria-producing rootlets that attach to, and sever, the host roots (Fig. 5.3). Seedlings can survive and grow without a host for up to 4 years, provided there are sufficient nutrients available in the soil.

Nuytsia produces numerous ring-like haustoria (called haustoriogens) that surround the host root. They have even been known to attach to underground telephone cables and trickle irrigation pipes (Fig. 5.3). Once the host root has been surrounded, a cutting device develops (Fig. 5.3) that commences to penetrate the root (Calladine & Pate, 2000). This cutting device is pushed through the host root by hydrostatic forces, and eventually becomes lodged in haustorial tissue on the opposite side of the host root. Endophytic tissue develops from the haustoriogen to each end of the host's cut xylem. There is no phloem in the haustorial endophyte, only paired xylem strands.

Fig. 5.3: (*upper left*)) *Nuytsia floribunda* growing on granite. Plants are 3 m tall, with the three visible stems arising from the same parent rhizome. (*upper right*) *Nuytsia* haustoriogens attached to an underground telephone cable (top) and *Banksia* host root (lower). Haustoriogens range from 5 to 10 mm in diameter. (*middle left*) Section through a haustoriogen that has severed a host root. The cutting device is pushed through the host root *via* hydrostatic pressure and in this photo is lodged in the left-hand side of the endophytic tissue. (*middle right*) The sharp sclerenchymatous horn (cutting device) that slices through the host root like a pair of shears, 4 mm long. (*lower left*) *Nuytsia* flowers occur in triads. From the single bisexual flower a winged fruit develops (*lower right*) held in place by three woody bracts, 15 mm long.

5.2.2 *Olax phyllanthi* (Olacaceae)

Olax is a root hemiparasite, with 5 species recorded as occurring in the SouthWest flora. The name *Olax* is derived from the Latin for "evil smelling" and refers to the unpleasant smell of some Asian species. *phyllanthi* relates to the pale soft leaves, that look similar to those of *Phyllanthus* (Euphorbiaceae).

O. phyllanthi is a shrubby species that establishes multiple haustorial contacts with roots of a wide range of woody and herbaceous host species (Pate *et al.*, 1990a; b). *Acacia littorea* on coastal dunes is a favoured natural host (Tennakoon *et al.*, 1997). Haustoria penetrate laterally to the host xylem where a predominately parenchymatous interface absorbs water, minerals and organic solutes from the host (Pate *et al.*, 1990a). *Olax* roots rarely reach below a depth of 30 cm, thus attachment to deep-rooted host species is essential for their summer survival. Haustorial contacts only form in autumn-spring and are made on both perennial and herbaceous species (Pate *et al.*, 1990b).

Damage to hosts is minimal and secondary thickening (of the host) continues on the side of a host root opposite to a haustorium. The developing sucker spreads laterally around the surface of the host xylem (Pate *et al.*, 1990b), but never completely encircles the host stele, unlike *Nuytsia.*

5.2.3 *Santalum* (Santalaceae)

Four species of *Santalum* occur in the SouthWest. The name is derived from the Greek *santalon*, meaning sandalwood (specifically referring to the Indian sandalwood). *S. acuminatum* (quandong) and *S. spicatum* (sandalwood) parasitize multiple woody host species, especially legumes such as *Acacia acuminata*.

Santalum has a 'bell-shaped' haustorium that partially encircles the host root (Fig. 5.4) and is extremely efficient at metabolising some of the amino acid (nitrogen-based) compounds transported in the host's transpiration stream. Like other hemiparasites, *Santalum* transpires more than its hosts (Tennakoon *et al.*, 1997; Loveys *et al.*, 2001).

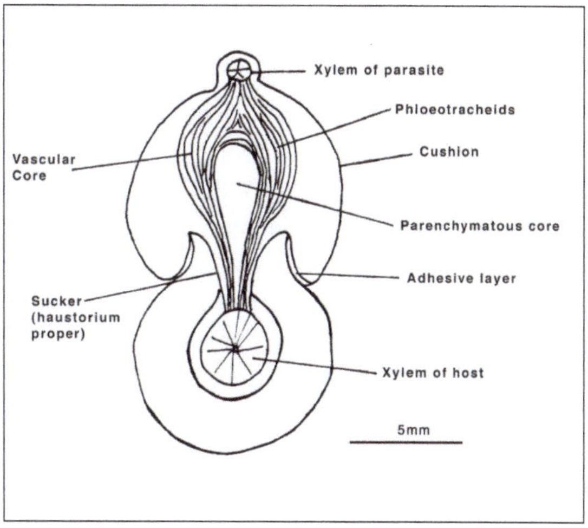

Fig. 5.4: (*left*) Leaves and fruits (3 cm in diameter) of *Santalum acuminatum* (quandong). (*right*) Drawing of longitudinal section through a *Santalum* haustorium attached to a host root.

5.3 Stem Holoparasites

5.3.1 *Pilostyles* (Apodanthaceae)

Pilostyles parasitizes only pea-flowered legume genera (Dell & Burbidge, 1981; Table 5.3). *Pilostyles* traditionally was positioned within the family Rafflesiaceae, which includes the holoparasitic, large-flowered *Rafflesia* (confined to southeast Asia), but is now considered a member of the related Apodanthaceae. *Pilostyles* species are small-flowered, with the name derived from the Greek words *pilos* (= hat) and *stylos* (= pillar) and refers to the shape of the style.

Table 5.3: Diagnostic features of the three SouthWest *Pilostyles* species (from Thiele *et al.*, 2008).

	P. hamiltonii	*P. coccoidea*	*P. collina*
Distribution	Eneabba to Bunbury	Eneabba to Moore River	Stirling Range, Hyden, Peak Charles
Host genus	*Daviesia*	*Jacksonia*	*Gastrolobium*
Sexual strategy	Dioecious	Monoecious	Monoecious
Host position	on young stems (< 2 year-old)	Usually on old wood; sometimes on young stems	On young stems

Legumes have the capacity, *via* a symbiotic association with nitrogen-fixing bacteria, to metabolise atmospheric nitrogen occurring in soil air pockets (see Chapter 6). *Pilostyles* targets legume species because legume tissues typically have a high nitrogen concentration. There are three *Pilostyles* species in the SW flora (Table 5.3). The plant body of *Pilostyles* is greatly reduced, and until it flowers, exists solely as a series of cellular strands within the secondary phloem of the host.

Pilostyles is completely dependent for its carbohydrate and inorganic nutrient requirements on uptake from phloem sieve tubes in the host stem. The phloem is part of a plant's vascular system that transports organic compounds, especially sucrose, and minerals up and down the stem (Fig. 5.5). *Pilostyles* strands occur predominantly in the phloem (Dell *et al.*, 1982). When reproductive tissues begin to form from the strands, *Pilostyles* starts to produce extensions ('pegs') into the host xylem. These pegs transfer extra water and minerals from the host xylem to the parasite. These extra nutrients are essential, as production of flower buds and flowers is a nutrient-expensive process.

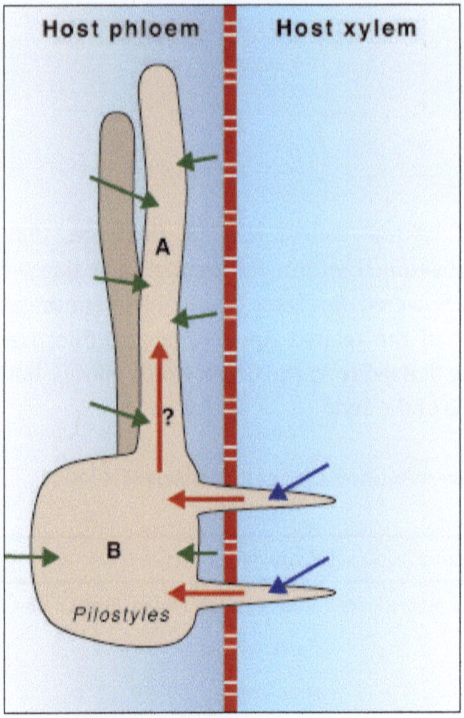

Fig. 5.5: Pathways of movement of host materials into *Pilostyles*. Vegetative tissue (A) utilizes primary phloem-derived nutrients, whereas reproductive tissues (B) utilize both phloem- (green arrows) and xylem-derived (blue arrows) nutrients. Red arrows represent internal parasite translocation of solutes. After Dell *et al.* (1982).

5.3.2 *Cuscuta* (Convolvulaceae)

This inconspicuous holoparasite is an annual herb with thread-like yellow stems that twine around host stems. Leaves are scale-like and the plant loses its roots once it reaches a host. Favoured hosts are in the daisy family (Asteraceae) and *Cuscuta* may sometimes be seen parasitizing other annuals, such as *Rhodanthe*. The scores of tiny haustoria penetrate the host's parenchyma, xylem and phloem (Lamont, 1982b). *Cuscuta campestris*, an exotic cosmopolitan species recorded in the SouthWest, may severely reduce the growth and flowering of its host (Shen *et al.*, 2005). Because of the non-green appearance of the plant and fact that it obtains carbohydrates from its host phloem it has usually been considered non-photosynthetic but chloroplasts have been detected. However, the activity of essential enzymes is neglible (Chudhury & Sahu, 1999). Only one native *Cuscuta* occurs in the SouthWest, *C. australis*.

5.4 Stem Hemi or Holoparasite? *Cassytha* (Lauraceae)

Cassytha is a scrambling, leafless, vine-like plant that is a stem parasite adhering to hosts *via* many minute haustoria. *Cassytha* is free-living, obtaining all its inorganic nutrient requirements from its host, and is considered to be an obligate parasite (Shen *et al.*, 2010). *Cassytha* species only tap host xylem, and provides its own source of carbon *via* photosynthesis. A seedling must attach itself to a host within a few weeks (McLuckie, 1924) otherwise it will perish. Once a host is parasitized the primary *Cassytha* stem dies back, losing contact with the soil.

A single *Cassytha* plant produces hundreds of haustoria, often on multiple hosts, and may also parasitize itself (autoparasitism) (Fig. 5.6). *Cassytha* is also known to display hyperparasitism, a situation where other parasitic species (e.g. *Santalum*, *Nuytsia*) are the hosts (Fig. 5.6). *Cassytha* haustoria are initiated before stem tissues are fully differentiated. A haustorium produces an adhesive disc that contains a thick layer of mostly unicellular, tightly packed microscopic hairs that fold at their tips upon contact with the host (Heide-Jorgensen, 1991). The hairs secrete adhesive mucilage that enables the haustorium to attach itself to the host. A wedge-shaped intrusive organ develops inside the layer of cells that lie immediately below the epidermis that penetrates the host stem *via* hydrostatic pressure and haustorial growth. Most of the starch that was stored in the adhesive disc before the intrusive phase is used during penetration (Fig. 5.6). The outer part of the storage tissue then forms a collapsed zone between the parasite and the host (Heide-Jorgensen, 1991).

Fig. 5.6: (*upper*) *Cassytha racemosa* (1–3 m wide) smothering host plants in a banksia woodland (*left*) and on the coastal dunes (*right*) near Perth, Western Australia. (*middle left*) *Cassytha* haustorial connections to a host stem 8 mm in diameter. (*middle right*) *Cassytha* hyperparasitising the root parasite *Nuytsia*. (*lower*) Sucker of *Cassytha* haustorium (on right) penetrating young stem of *Acacia saligna* (on left) until it connects to the host's xylem. Note starch grains in the 0.5 mm-wide core of the haustorium that serve to meet the energy requirements of the dynamic processes involved.

6 Specialised Nutrient Uptake Mechanisms

Nitrogen (N) and phosphorus (P) are major nutrients for plant growth and function. In the nutrient-impoverished soils of the SouthWest, both N and P need to be added to the soil for satisfactory growth of crop plants. N can be assimilated from atmospheric N_2 by microorganisms and is highly mobile as inorganic N in the soil. Soil P is a highly immobile, non-renewable resource. Most inorganic P is adsorbed by humus and clay particles, rendering it unavailable for plants, unless they have specialised roots for solubilising adsorbed P.

6.1 Nitrogen

Plants require N to produce amino acids, proteins, enzymes and nucleic acids. N is typically acquired from the soil in the form of nitrate ions (NO_3^-). When growing in N-impoverished soils, plants have adapted to utilise the available N very efficiently. N is abundant in the atmosphere (almost 80% N_2) and can be fixed into N-based organic compounds. The capacity to fix atmospheric nitrogen is undertaken by microorganisms possessing the enzyme nitrogenase, the only enzyme capable of reducing N_2 to ammonia (NH_3) for easy conversion into ammonium (NH_4^+). Ammonium becomes incorporated by N_2-fixing organisms into the amino acid, glutamate, and ultimately into proteins and other N-containing molecules. These microorganisms are known to form symbiotic associations with vascular plants and some lichens, providing them with a source of organic N. This process occurs most efficiently under oxygen-free conditions as N_2-fixation is an anaerobic process.

6.1.1 Cyanobacterial Symbionts

Cycads are gymnosperms that were major component of the world's flora during the Permian–Triassic and arose about 330 Ma. They have a symbiotic relationship with cyanobacteria (usually *Nostoc*) sometimes referred to as blue-green algae. In the SouthWest, cycads are represented by three species of *Macrozamia*. The structures in which nitrogen fixation occurs in cycads are called coralloid roots, so named because they are coral-like in appearance (Fig. 6.1). The nitrogen-fixing cyanobacteria are located within the middle cortex of the coralloid root (Fig. 6.1), and use the carbon-based energy sources supplied by the host root (Halliday & Pate, 1976). The densely packed cyanobacteria occupy the spaces between the cells and force them to elongate. Coralloid roots are attached to the parent tap root up to 50 cm below the soil surface. Within 12 months of germination, cycad seedlings produce special lateral roots at the root-shoot junction called precoralloid roots (Lamont & Ryan, 1977). Eventually they are transformed into coralloid roots, once colonised by cyanobacteria. Some lichens, e.g. *Leptogium*, have cyanobacteria as symbionts with their fungal partner, rather than the more usual green algae (e.g. the common SouthWest lichen *Parmelia* that occurs on granite outcrops).

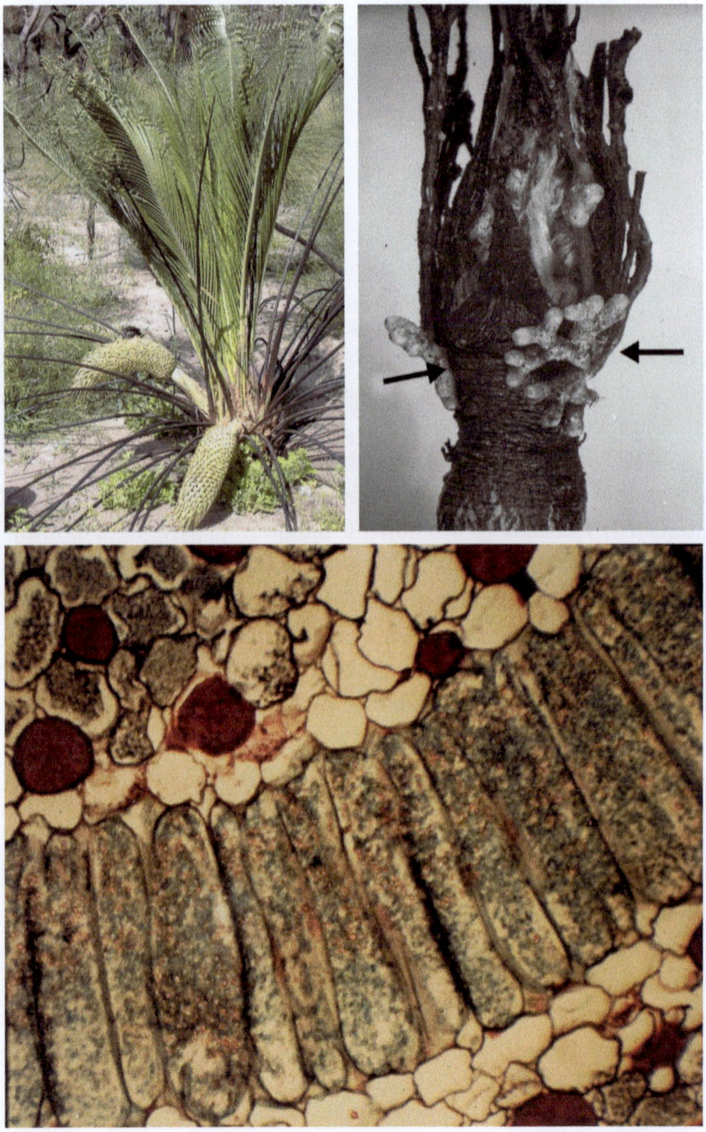

Fig. 6.1: (*upper left*) The cycad, *Macrozamia riedlei*, producing cones after a fire, utilising extra nitrogen supplied by enhanced nitrogen fixation following fire. Fronds 1.5 m long. (*upper right*) Coralloid roots (arrowed) of a young *Macrozamia riedlei*. (*lower*) transverse section through coralloid root of *M. riedlei* showing cyanobacteria packed within a single layer of the root cortex. Width of algal zone is 0.2 mm.

6.1.2 Rhizobial Symbionts

Nitrogen-fixing bacteria known as 'rhizobia' (after the genus *Rhizobium*) have a symbiotic relationship with legumes (Fabaceae), except subfamily Caesalpinioideae. Legume nodulation originated 58 Ma, within 2 My after the appearance of legumes (Sprent, 2007). Nitrogen-fixing bacteria are present in vast numbers within infected root cells, surrounded by a protective membrane provided by the host root and are termed bacteriods. Bacteroids are located within the central cortex of the legume root nodule (Fig. 6.2). The outer layer of nodule cells acts as a barrier to the inward diffusion of oxygen from the soil. The inner nodule cells produce large quantities of the oxygen-binding protein known as leghaemoglobin that delivers oxygen directly to the bacteroids, sustaining respiration and also ensuring an oxygen-free environment for nitrogen fixation to occur. The 'haem' component is produced by the bacteria, and the 'globin' component is produced by the plant. Root nodules tend to have a pinkish-brown colour due to the presence of leghaemoglobin.

6.1.3 Actinorhizal Symbionts

Casuarina and *Allocasuarina* (Casuarinaceae) often possess root nodules (Fig. 6.1) induced by infection from filamentous Actinomycete soil bacteria (or actinobacteria) in the genus *Frankia* (Lamont, 1982; Pate, 2002). *Frankia* isolates from *Casuarina* and *Allocasuarina* nodules are from similar strains but differ from those in the related *Gymnostoma* (Navarro, Rouvier, Normand, *et al.*, 1998) that reflects the early divergence of *Gymnostoma* within the Casuarinaceae (Crisp *et al.*, 2004). Actinorhizal nodules are clusters of branching rootlets inhabited by *Frankia* bacteria and can exceed 5 cm in diameter. Casuarinales is the only group of actinorhizal plants worldwide where the nitrogen fixation process does not occur within specialised microbial vesicles, and the nodules also lack the haemoglobin pigment essential for N_2 fixation in legumes. In waterlogged habitats nodules may be located close to the surface, whereas, in drier habitats, the nodules are deeper in the soil profile.

Fig. 6.2: (*upper*) Cluster of nodules on a *Swainsona formosa* (Fabaceae) root and *S. formosa* flowers. Cluster is 2 cm wide, flowers are 5 cm long. (*middle) Paraserianthes lophantha* (Fabaceae) colonizing a creek bed in the Darling Range showing part of a 4-m-tall plant, and a longitudinal section through one of its root nodules showing the masses of bacteroids (stained red) in the middle cortex and vascular tissue in the outer cortex. (*lower*) Cluster of actinorhizal root nodules, 4 cm wide, of *Casuarina obesa* that occurs in naturally anaerobic waterlogged soils along watercourses in the SouthWest. (*right*) *Allocasuarina fraseriana*, 5 m tall, though nodules have yet to be found on this species.

6.2 Phosphorus

Phosphorus is the macronutrient that is most severely limited in SouthWest soils. Plants require P as it forms an important component of the genetic code (in DNA and RNA), is a structural component of cell membranes (phospholipids), and is an important part of the energy-carrying molecule ATP. The most common responses to low P in the SouthWest flora, mycorrhizal associations and root clusters, are examined in this chapter. There are various methods that plants have enlisted to enhance their P uptake in P-impoverished soils:

1. Increased root-hair production (increased surface area) and penetration of spaces between soil particles.
2. Increased production of rootlets (increased exudation of organic acids and enzymes into the rhizosphere around the roots).
3. Increased infection by mycorrhizal fungi whose fine hyphae penetrate soil spaces and absorb P preferentially.
4. Increased efficiency of P uptake by adhering to soil particles *via* mucigel and active transport.

6.2.1 Mycorrhizal Associations

The term mycorrhiza comes from the Greek *mycors*, meaning fungus (literally toadstool) and *rhiza*, meaning root, and is a mutualistic symbiotic association between fungal hyphae and a plant root. The fungus receives carbohydrates and other organic compounds from the host plant in exchange for functioning as an extended root system, thereby greately improving mineral uptake by the plant roots (Brundett, 2004).

About 80% of all terrestrial plants have some sort of mycorrhizal association (Brundrett, 2009). Arbuscular mycorrhizas (AM) are the most abundant in the SouthWest flora (51%), followed by ectomycorrhizas (EM) (20%). Both occur on actively growing or recently formed roots. SouthWest species that have dual EM/AM associations occur within the Casuarinaceae, Fabaceae, Myrtaceae and Rhamnaceae. Other mycorrhizal types exist and display a greater degree of host specifity, e.g. orchid mycorrhizas and ericoid mycorrhizas. Orchid mycorrhizas differ from other mycorrhizal types because they occur in roots and tubers of the Orchidaceae. Ericoid mycorrhizas infect hairless roots within the Ericaceae.

6.2.2 Arbuscular Mycorrhizas

Most mycorrhizal associations are arbuscular mycorrhizas (AM). The fungal component is always in the fungal phylum Glomeromycota. AM species are not host specific, and a 5 cm length of living root can have as many as eight species from five AM genera. The host plant may be facultatively or obligately dependent on its fungal partner. AM are endomycorrhizas as the fungal strands invade the host root cells (Fig. 6.3). Once inside the cortex they form tree-shaped structures called arbuscules that often fill the cells. Because the arbuscular hypae are so finely divided and surrounded by the host cell membrane, both symbiotic partners are brought into very close contact. This highly specialized symbiosis was previously known as "vesicular-arbuscular mycorrhizas" (VAM) because some glomeromycotan fungi also form storage hyphae termed vesicles within the root cells.

Mycorrhizas enhance the surface area available for uptake, especially important for ions that diffuse slowly in soil such as P. AM enhance inorganic P uptake by increasing the surface area available for absorption, and are able to grow an extensive network of fine hyphal strands, one-tenth the diameter of root hairs, into parts of the soil that the root hairs cannot reach (Fig. 6.3). Uninfected roots create a P-depletion zone caused by high P uptake and low P diffusion rates. The mycelium of AM grows far beyond the depletion zone, reaching a new pool of soluble P. Some types of AM acquire inorganic N from organic sources among the humus particles.

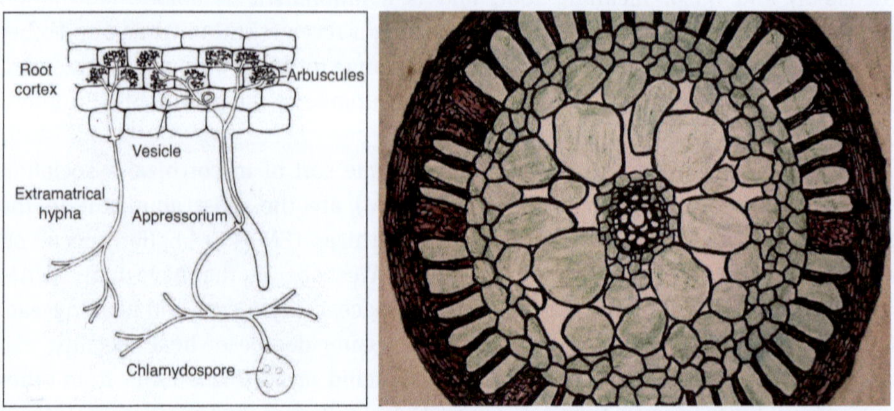

Fig. 6.3: (*left*) AM hyphae inhabiting the root cortical and root hair cells of *Eucalyptus gomphocephala.* Note the presence of arbuscules and a storage vesicle. Cortical cells about 100 µm wide. (*right*) Ectomycorrhizal sheath surrounding a *Melaleuca squarrosa* (Myrtaceae) root. The hyphae within the root are located between the epidermal cells in a formation known as the Hartig net, and so do not infect living cells. Roolet is 1 mm in diameter.

6.2.3 Ectomycorrhizas

Ectomycorrhizas (EM) occur predominantly within the fungal phylum Basidiomycota (that contains the mushroom-like fungi and puffballs), and to a lesser extent within the phylum Ascomycota and the 'Zygomycetes'. EM form a hyphal sheath up to 30 µm wide around rootlets of the host plant (Fig. 6.3). EM are host-specific, and may synthesise substances that cause the morphology of the root to change. The sheath (or mantle) formed over the root surface suppresses root hair development (Fig. 6.3), and all nutrients and water absorbed by the affected area have to pass through the fungal sheath. Hyphae arising from the mantle penetrate the host root between the epidermal cells and it is here where transfer of nutrients occurs. EMs are sometimes referred to as ectotrophic associations or sheathing mycorrhizas. Like AM, EM are able to access new pools of soil P (and other nutrients) unavailable to uninfected roots because their hyphae are able to penetrate into parts of the soil that are not occupied by the host roots. Table 6.1 lists species with mycorrhizal and non-mycorrhizal roots present in Western Australia's jarrah (*Eucalyptus marginata*) forest. Note that some species can possess both AM and EM, whereas others are mycorrhizal-type specific.

The critically endangered achlorophyllous subterranean orchid, *Rhizanthella gardneri*, is an obligate mycoheterotroph, meaning that it is entirely dependent on its mycorrhizal association to meet both its carbon and mineral nutrient requirements (Hyson *et al.*, 2013). *R. gardneri* is restricted to seasonally-damp habitats dominated by species in the *Melaleuca uncinata* complex (Bougoure *et al.*, 2008) and forms a tripartite relationship involving the in-common EM (Bougoure *et al.*, 2009), whereby *Melaleuca* provides carbon-based compounds and nutrients *via* the mychorrizal bridge between the shrub and the underground orchid (Bougoure *et al.*, 2010) and from the fungi's saprophytic soil acitivity (Fig. 6.4). It is unlikely that the orchid provides the host shrub or fungus with any nutrients in return, so that it is actually an example of an indirect holoparasitic relationship with an autotrophic partner.

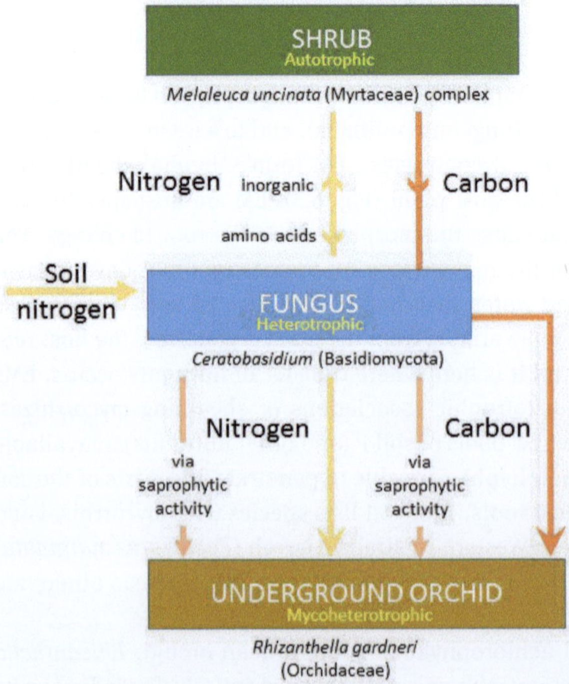

Fig. 6.4: Tripartite relationship between the underground orchid (*Rhizanthella gardneri*), its fungal symbiont, *Ceratobasidium* (Basidiomycota: Cantharellaes), and shrubs in the *Melaleuca uncinata* complex. *R. gardneri* is achlorophyllous and an obligate mycoheterotroph, with hyphal strands of *Ceratobasidium* providing a continuum between orchid and shrub. *Ceratobasidium* species are ectomycorrhizal on the *Melaleuca* roots and form pelotons in the orchid's rhizome. The orchid obtains carbon produced by the autotrophic shrub host *via* its fungal connections. It also receives carbon (and nitrogen) *via* the saprophytic litter-decomposing activities of the fungus. The fungus absorbs soil nitrogen that is transported to the shrub and orchid. Based upon the work of Bougoure *et al.*, 2009 and Bougoure *et al.*, 2010.

6.2.4 Orchid and Ericoid Mycorrhizas

Orchid mycorrhizas are endophytic (occuring within root cells), representing a symbiotic relationship with fungi in the Basidiomycota or Ascomycota. For some, or all, of their life cycle, orchids are obligately dependent on mycorrhizas. They are essential for germination as orchid seeds are tiny and lack an endosperm for providing carbohydrates needed for early growth. This is not a mutualistic association as, in most cases, the mycorrhiza receives no benefit in return (Brundrett, 2007).

Ericoid mycorrhizas are mainly in the Ascomycota and are a feature of the Ericaceae family. Also known as 'hair roots', ericoid mycorrhizas exist as branched coils of fungal hyphae within the outer cortical cells of hairless rootlets of ericaceous species and are known to absorb organic N (Dixon *et al.*, 2002).

Table 6.1: Presence of arbuscular mycorrhizas (AM), ectomycorrhizas (EM), other types of mycorrhizal and non-mycorrhizal roots (NM) in plant species inhabiting the northern jarrah forest (55 km SE of Perth). Data obtained from Brundrett & Abbott (1991) with species names updated to represent current taxonomy.

Taxa	Growth form	Mycorrhizal type	Other mechanisms
Zamiaceae			
Macrozamia riedlei	Palm-like	AM	Coralloid roots
Ranunculaceae			
Clematis pubescens	Climber	AM	
Casuarinaceae			
Allocasuarina humilis	Shrub	EM/AM	*Frankia* nodules
A. fraseriana	Tree	EM/AM	(not all Casuarinaceae)
Dilleniaceae			
Hibbertia amplexicaulis	Subshrub	AM	
H. commutata	Subshrub	AM	
Malvaceae			
Lasiopetalum bracteatum	Subshrub	AM	Beaded roots
Thomasia glutinosa	Subshrub	AM	
Droseraceae	Geophytes:		
Drosera erythrorhiza	Rosette	NM	Carnivorous
D. macrantha	Climbing	NM	
D. platystigma	Pygmy	NM	
Ericaceae			
Leucopogon verticillatus	Shrub	Ericoid	
Pittosporaceae			
Billardiera heterophylla	Climber	AM	
Marianthus bicolor	Climber	AM	
Fabaceae (Mimosoideae)			
Acacia alata	Subshrub	AM	Rhizobial nodules
A. extensa	Shrub	AM	
A. lateriticola	Shrub	AM	
A. pulchella	Shrub	AM	
A. saligna	Shrub	AM	
A. urophylla	Shrub	AM	
Fabaceae (Caesalpinioideae)			
Labichea punctata	Shrub	AM	

continued **Table 6.1:** Presence of arbuscular mycorrhizas (AM), ectomycorrhizas (EM), other types of mycorrhizal and non-mycorrhizal roots (NM) in plant species inhabiting the northern jarrah forest (55 km SE of Perth). Data obtained from Brundrett & Abbott (1991) with species names updated to represent current taxonomy.

Taxa	Growth form	Mycorrhizal type	Other mechanisms
Fabaceae (Faboideae)			Rhizobial nodules on all Faboideae
Bossiaea aquifolium	Subshrub	AM	Root clusters
B. ornata	Shrub	AM	
Daviesia cordata	Shrub	NM	Root clusters
D. decurrens	Shrub	NM	Root clusters
Gastrolobium bilobum	Shrub	EM/AM	
G. capitatum	Shrub	EM/AM	
G. lanceolatum	Shrub	EM/AM	
Gompholobium marginatum	Shrub	EM/AM	
G. tomentosum	Shrub	EM/AM	
G. knightianum	Subshrub	EM/AM	
Hardenbergia comptoniana	Climber	AM	
Hovea chorizemifolia	Shrub	AM	
H. trisperma	Shrub	AM	
Jacksonia floribunda	Shrub	AM	
Kennedia prostrata	Creeper	AM	Root clusters
K. coccinea	Climber	NM	
K. rubicunda	Climber	AM	
Mirbelia dilatata	Shrub	EM/AM	
Viminaria juncea	Shrub	AM	Root clusters
Proteaceae			Proteoid root clusters on all SouthWest Proteaceae, except *Persoonia* and *Acidonia*
Adenanthos barbiger	Shrub	NM	
Banksia grandis	Tree	NM	
B. dallanneyi	Shrub	NM	
B. sphaerocarpa	Shrub	NM	
Conospermum capitatum	Subshrub	NM	
Grevillea bipinnatifida	Shrub	NM	
G. quercifolia	Shrub	NM	
G. wilsonii	Shrub	NM	
Hakea ceratophylla	Shrub	NM	
H. lissocarpha	Shrub	NM	
Isopogon dubius	Shrub	NM	
Stirlingia latifolia	Shrub	NM	
Synaphea petiolaris	Subshrub	NM	
Xylomelum occidentale	Tree	NM	
Thymelaeaceae			
Pimelea imbricata	Subshrub	AM	

continued **Table 6.1:** Presence of arbuscular mycorrhizas (AM), ectomycorrhizas (EM), other types of mycorrhizal and non-mycorrhizal roots (NM) in plant species inhabiting the northern jarrah forest (55 km SE of Perth). Data obtained from Brundrett & Abbott (1991) with species names updated to represent current taxonomy.

Taxa	**Growth form**	**Mycorrhizal type**	**Other mechanisms**
Myrtaceae			
Agonis flexuosa	Tree	EM/AM	
Calytrix leschenaultii	Subshrub	AM	
Corymbia calophylla	Tree	EM/AM	
Eucalyptus drummondii	Tree	EM/AM	
E. marginata	Tree	EM/AM	
E. patens	Tree	EM/AM	
Hypocalymma angustifolium	Shrub	EM/AM	
Melaleuca lateritia	Shrub	EM/AM	
M. scabra	Shrub	EM/AM	
Pericalymma ellipticum	Shrub	EM/AM	
Taxandria linearifolia	Shrub	EM/AM	
Verticordia plumosa	Subshrub	AM	
Loranthaceae			
Nuytsia floribunda	Tree	NM	Parasitic
Celastraceae			
Tripterococcus brunonis	Herb	AM	
Euphorbiaceae			
Phyllanthus calycinus	Shrub	AM	
Rhamnaceae			
Cryptandra arbutiflora	Shrub	EM/AM	
Trymalium floribundum	Shrub	EM/AM	
Tremandraceae			
Tetratheca hirsuta	Subshrub	AM	
Polygalaceae			
Comesperma virgatum	Subshrub	AM	
Rutaceae			
Boronia spathulata	Shrub	AM	
Philotheca spicata	Shrub	AM	
Apiaceae			
Eryngium rostratum	Herb	AM	
Trachymene pilosa	Annual	AM	
Xanthosia candida	Herb	AM	
Lamiaceae			
Hemigenia sericea	Subshrub	AM	
Lobeliaceae			
Lobelia tenuior	Herb	AM	

continued **Table 6.1:** Presence of arbuscular mycorrhizas (AM), ectomycorrhizas (EM), other types of mycorrhizal and non-mycorrhizal roots (NM) in plant species inhabiting the northern jarrah forest (55 km SE of Perth). Data obtained from Brundrett & Abbott (1991) with species names updated to represent current taxonomy.

Taxa	Growth form	Mycorrhizal type	Other mechanisms
Stylidiaceae			
Stylidium brunonianum	Herb	AM	
S. junceum	Herb	AM	
S. piliferum	Herb	AM	
Goodeniaceae			
Dampiera linearis	Herb	AM	
Lechenaultia biloba	Herb	AM	
L. floribunda	Herb	AM	
Scaevola calliptera	Herb	AM	
Rubiaceae			
Opercularia echinocephala	Shrub	AM	
Asteraceae			
Hyalosperma cotula	Annual	AM	
Olearia paucidentata	Shrub	AM	
Waitzia suaveolens	Annual	AM	
Dasypogonaceae			
Kingia australis	Grasstree	NM	Long root hairs, aerial roots embedded in leafbases
Xanthorrhoeaceae			
Xanthorrhoea gracilis	Subshrub	AM	
X. preissii	Grasstree	AM	
Asparagaceae			
Chamaescilla corymbosa	Herb	AM	
Thysanotus dichotomus	Herb	Other	
T. multiflorus	Herb	Other	
Hemerocallidaceae			
Agrostocrinum scabrum	Herb	AM	
Colchicaceae			
Burchardia congesta	Herb	AM	
Iridaceae			
Patersonia occidentalis	Herb	AM	
Orchidaceae			
Thelymitra crinita	geophyte	Orchid	

continued Table 6.1: Presence of arbuscular mycorrhizas (AM), ectomycorrhizas (EM), other types of mycorrhizal and non-mycorrhizal roots (NM) in plant species inhabiting the northern jarrah forest (55 km SE of Perth). Data obtained from Brundrett & Abbott (1991) with species names updated to represent current taxonomy.

Taxa	Growth form	Mycorrhizal type	Other mechanisms
Haemodoraceae			
Anigozanthos manglesii	Herb	NM	
Conostylis setosa	Herb	NM	
Haemodorum laxum	Herb	NM	
H. spicatum	Herb	NM	
Restionaceae			
Desmocladus flexuosus	Herb	NM	Long root hairs and very fine roots
Poaceae			
Tetrarrhena laevis	Herb	AM	

6.3 Case Study: Dispersal of Ectomycorrhizal Fungi by Marsupials

The woylie, *Bettongia penicillata* (Potorodiae), is a small marsupial whose diet includes the underground fruiting bodies (sporocarps) of at least 18 species of mushrooms or puffballs in SouthWest eucalypt forests. These mycophagous (fungus-eating) mammals are important dispersers of ectomycorrhizal fungal spores, particularly in the puffball genus *Mesophellia* (Basidiomycota: Mesophelliaceae).

Experiments with seedlings of the legume *Gastrolobium bilobum* and the eucalypt *Corymbia calophylla*, using woylie faecal pellets (Lamont *et al.*, 1985), have shown that more ectomycorrhizal roots are formed when fresh woylie faecal pellets are applied to the soil than freshly collected spores of *Mesophellia*. This implies that woylie faecal material is a reliable source of viable spores, and that digestion provides a pre-treatment for spore germination. When autoclaved pellets were used, no mycorrhizas formed and plants were one-tenth the weight of those with EM produced *via* spores from the pellets.

This demonstrates that a triangular trophic relationship exists (Fig. 6.5). These plant species are dependent on mycorrhizas for nutrient uptake, in turn providing shelter for the woylie, an important disperser of viable, pre-treated fungal spores that infect the plant rootlets. The role of mycophagous marsupials is particularly important for dispersal of these EM fungi whose spores cannot become airborne because their fruiting bodies only occur below the soil surface (Claridge, 2002). Bettongs arose 28 Ma (Prideaux & Warburton, 2010) and may have had a significant role in the evolution of ectomycorrhizal shrubs and trees in the SouthWest. Loss of mycophagous marsupials is having a deterimental effect on the functioning of Australian forest ecosystems throughout Australia as nutrient cycling is now curtailed (Lamont, 1992; Fleming *et al.*, 2014). Native dung beetles may also be part of the triangle.

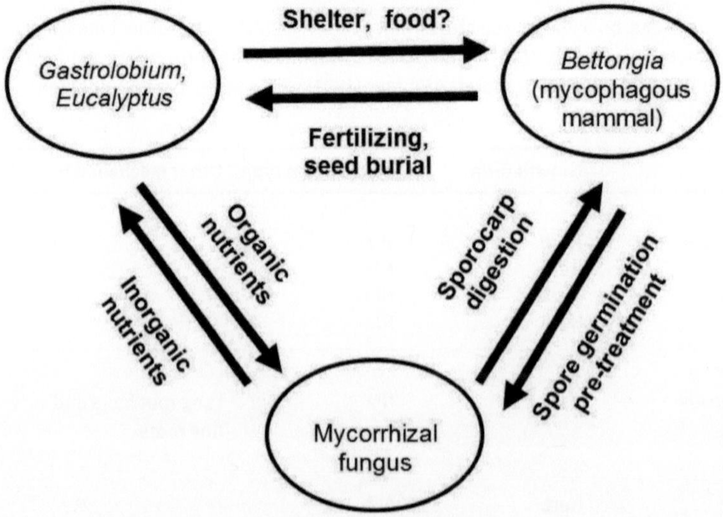

Fig. 6.5: Tripartite relationship between a mycophagous marsupial (*Bettongia*), ectomycorrhizal fungi and their host plant species. Adapted from Lamont *et al.*, 1985.

6.4 Root Clusters

The term 'root cluster' (or cluster root) is a generic term used to define a number of bottlebrush-like branched root structures (Lamont, 2003). These include proteoid root clusters in most of the Proteaceae and some Fabaceae, dauciform root clusters in sedges (Cyperaceae), and capillaroid roots in the Restionaceae. Cluster roots are produced in response to low soil fertility and, by creating a surface area substantially greater than 'normal' roots, have the primary function of providing greater access to stores of soil P and micronutrients. Thus they function in a similar way to mycorrhizas.

6.4.1 Proteoid Roots

Proteoid roots (Fig. 6.6) are a characteristic feature of the Proteaceae (except *Persoonia*) and are concentrated in the organic-rich layer of the soil, often forming large surface root mats, 5-10 cm thick and growing up into the decomposing litter (Lamont, 2003). Proteoid roots resemble 'bottlebrushes' and can be classified as simple (e.g. *Hakea*) or compound (e.g. *Banksia*). Compound clusters are a collection of simple clusters. Within the SouthWest flora, proteoid roots are also well developed in the legume *Viminaria juncea*, a species confined to winter-wet depressions (Fig. 6.6). In the Australian flora, proteoid roots also occur in some *Casuarina* and legume species, such as *Kennedia prostrata*.

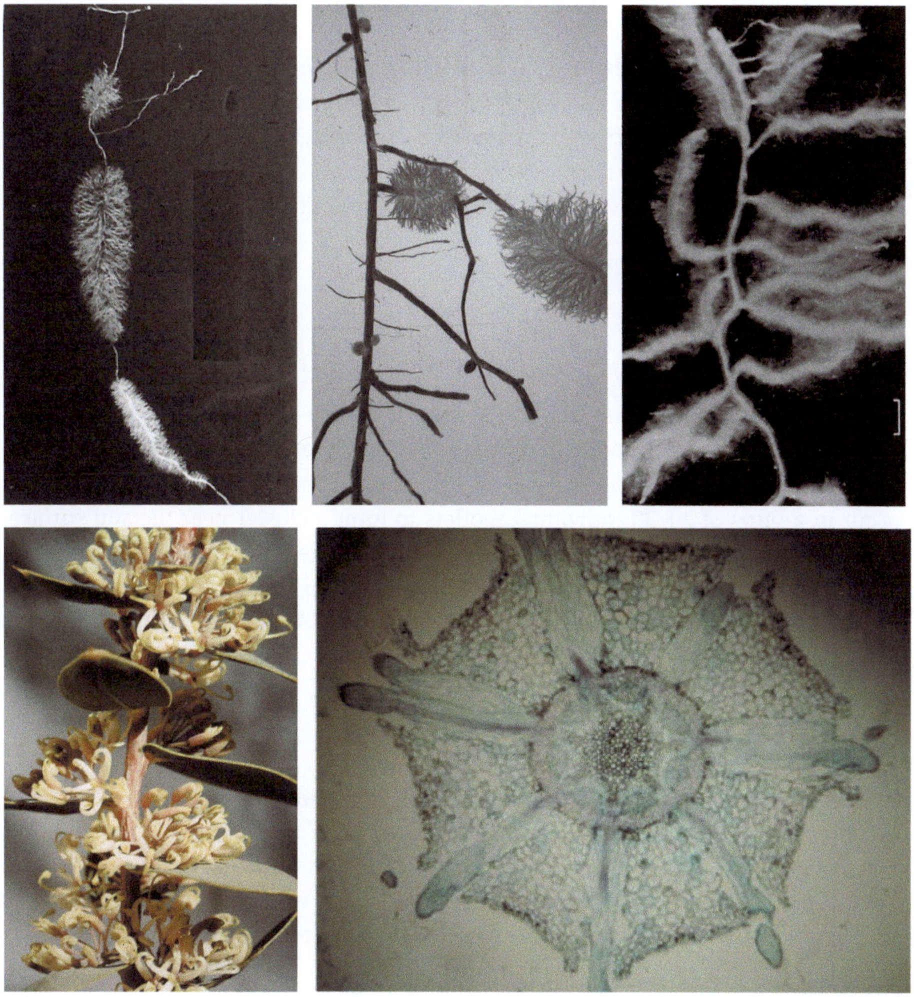

Fig. 6.6: Proteoid roots of (*upper left*) 3-month-old *Hakea lissocarpha* (Proteaceae) root, displaying successive formation of root clusters – from recently formed (lower) to senescent (upper). Middle cluster is 8 cm long. (*upper middle*) *Viminaria juncea* (Fabaceae) showing two clusters, as well as root nodules and pneumatophores (roots for aeration that point up into the water when flooded). Larger cluster is 5 cm long. (*upper right*) A cluster of dauciform roots on the sedge, *Cyathochaeta avenacea* (Cyperaceae). Scale = 1 mm. (*lower left*) *Hakea prostrata*, widespread through the SouthWest, with axillary flower clusters protected from florivores by rigid 4-cm-long leaves with a sharp apex. 2-mm-wide transverse section of one of its young proteoid roots (*lower right*, stained with safranin and fast green), with the rootlets arising in seven rows from the stele and often in pairs, explaining how it may have up to 1,000 rootlets per cm length of parent root.

Extensive work on the function and ecological significance of proteoid roots has shown that the soil immediately adjacent to these clusters is slightly more acidic than the surrounding soil (Lambers *et al.*, 2006). This is because the clusters produce organic acid exudates (especially citric acid and lesser extent, malic acid), and phosphatase enzymes that release P that is otherwise unavailable to the plant. Other highly immobile nutrients such as iron and manganese are also oxidised into an ionic form that can be taken up by the root system. It has recently been shown that soil bacteria may have a key role in the formation of proteoid root clusters (Lamont *et al.*, 2015).

6.4.2 Dauciform Roots

Dauciform roots (Fig. 6.6) are characterised by relatively short (2–12 mm long), carrot-shaped (*daucus* = carrot) rootlets covered with densely packed ~2 mm-long root hairs arising from elongated epidermal cells (Lamont, 1974; Shane *et al.*, 2005). These often occur in groups of 20–30 individual rootlets to form a cluster near the soil surface. Dauciform roots occur in some species within certain genera of the Cyperaceae (e.g. *Carex*, *Lepidosperma* and *Schoenus*) (Shane *et al.*, 2005; Barrett, 2013), and function in a similar way to proteoid roots (Playsted *et al.*, 2006; Shane *et al.*, 2006).

6.4.3 Capillaroid Roots

Capillaroid roots are characterised by clusters of rootlets, densely covered with exceptionally long root hairs that form a mat at the soil surface (Lamont, 1982b). The name reflects their sponge-like properties in retaining soil water. Little is known about the functional significance of capillaroid roots within the Restionaceae but they appear to be analogous to proteoid and dauciform roots.

6.5 The Bizarre Root System of *Kingia australis*

Kingia australis (Dasypogonaceae) is a grasstree that occurs on the margins of wetlands along coastal parts of the SouthWest forest region and drier slopes of the Gairdner Range in the northern sandplains. They can be mistaken for 'blackboys' (*Xanthorrhoea*) except that plants are slimmer and taller (up to 10 m) and have a crown of 'drumstick' flower spikes that form best in response to fire (Fig. 6.7; Lamont & Downes, 1979). Removing the persistent leaf bases reveals a mantle of hidden aerial roots that arises from near the apex of the true stem and forms the major part of the caudex at ground level (Lamont, 1981a). The roots grow at 15 cm per year taking 35 years to reach the ground in a 6-m-tall kingia and there may be 3000 aerial roots entering the soil at that point. The underground root system is composed entirely of these roots that may individually reach 18 m in length and survive for 300–400 years. Within the caudex, these primary roots produce lateral branches that spread among the leaf bases and possess root hairs 2 or more mm long.

Fig. 6.7: The root system of *Kingia australis*. (*upper left*) Two plants 1.8 and 2.3 m tall with crowns of flower spikes. Undulations on the persistent leaf bases correspond to annual growth increments (Lamont & Downes, 1979). (*upper right*) Leaf bases removed to reveal rings of aerial roots running down the stem from near the apex. In young plants they are produced annually but in older plants their production is stimulated by fire (Lamont, 1981a). Section is 10 cm wide. (*lower left*) Transverse section of caudex 20 cm above ground to show ring of root primordia on the 6-cm-wide stem surrounded by hundreds of aerial roots pushing through the leaf bases. *(lower right)* lateral roots proliferate in the spaces between leaf bases and may sometimes be seen protruding beyond them. Squares are 1 cm.

The aerial roots provide mechanical support for the stem in the absence of secondary thickening that occurs in xanthorrhoeas and their habitats are far too fire- and drought-prone for the formation of equivalent prop roots possessed by some tropical species with thin stems, such as *Pandanus*. In addition to the protection offered by the leaf bases, they can act as a source of water and nutrients for absorption by the laterals. Sources of nutrients include the decaying leaf bases themselves and dust particles, insect bodies, bird droppings (the spikes are visited by birds) and ash caught in the leaf bases when the leaves burn–the levels of K, N and P are 2–7 times that in an equivalent volume of soil around the plant (Lamont, 1981b). Labelled P injected among the leaf bases half way along the stem to simulate such a source is rapidly translocated up the primary roots into the true stem and from there to all other physiologically active parts of the plant (Lamont, 1984b).

6.6 Nutrient-impoverished Soils Promote Species Richness

Worldwide, P-impoverished soils support the greatest plant species richness (Lambers *et al.*, 2010; Venterink, 2011; Blanck *et al.*, 2011), associated with a array of complex strategies not only to maximise P uptake (this chapter), but also to ensure that P is used efficiently and effectively (e.g. increased leaf longevity, see Chapter 9) and P-enriched seeds (see Chapter 11). Phosphorus becomes increasingly limited as soil ages (Lambers *et al.*, 2008), and SouthWest soils are extremely old and weathered (Hopper, 2009; Mucina & Wardell-Johnson, 2011),

The hyperdiverse SouthWest sandplains possess the lowest soil P concentrations of all the world's Mediterranean-type ecosystems (Lamont, 1995; Cowling *et al.*, 1996), the greatest diversity of P uptake strategies (Lamont, 1982b; Lambers *et al.*, 2010) and high species and phylogenetic diversity (Lamont *et al.*, 1977), fine-scale turn over and levels of local endemism (Lamont *et al.*, 1984; Merwin *et al.*, 2012; Mucina *et al.*, 2014). Phosphorus conservation and exploitation promote diversification through increased competition for limited soil resources at the phylogenetic and species levels. In the SouthWest sandplains P-impoverished soils are also associated with low levels of N and K (Lamont, 1995) that has enabled the radiation and form diversficiation of carnivorous *Drosera* species to the point that the sandplains are a biodiversity centre for sundews. Indeed, the SouthWest nutrient-impoverished soils are the global centre for a vast array of families and genera, including many that are monotypic (Lamont *et al.*, 1984; Mucina *et al.*, 2014).

7 Pollination Strategies

Pollination is the transfer of pollen from an anther to the receptive surface of a stigma within the same species. Most species are hermaphroditic (also known as bisexual) where both male and female parts occur within the same flower. One of the most efficient breeding mechanisms is selfing (autogamy) or self-pollination whereby pollen is accepted by the same flower. Where it does not require a pollinator it is called spontaneous autogamy. The disadvantage is that selfing promotes inbreeding and does not allow for novel combinations of genes as a basis for adaptation to habitat change and speciation. In addition, inbreeding may lead to a decline in plant vigour due to expression of less well-adapted recessive and deleterious genes. Thus, inbreeding may ultimately lead to declining plant vigour within the population. Hence, most species are outbreeders (xenogamy).

One way of guaranteeing outbreeding is to have unisexual flowers on separate plants (dioecy). About 5% of southwestern Australian species are dioecious (e.g. *Wurmbea dioica* (Colchicaceae); Case & Barrett, 2004). Another method is to have male- and female-only flowers on the same plant (monoecy) that accounts for about 3% of SouthWest species (e.g. *Diplopeltis*, Fig. 7.1) (McComb, 1966). Monoecy only works as an outbreeding device if the opening of the different flower types on a plant is out of phase, as it usually is.

Monoecious plants may have hermaphroditic flowers as well as male or female flowers on the same plant. When hermaphroditic and male flowers are present this is known as andromonoecy that occurs within the Proteaceae (e.g. *Stirlingia*, Ladd & Connell, 1994) and the Myrtaceae (e.g. *Melaleuca, Beaufortia*) (Beardsell *et al.*, 1993). *Nuytsia floribunda* (Loranthaceae), the iconic Western Australian christmas tree, is also andromonoecious. Flowers occur in triads, with a central hermaphroditic flower nestled between two functionally male flowers. Thus only one flower sets fruit that sits with its three wings between the three bracteoles that elongate after seed set. When female and hermaphroditic flowers are present it is known as gynomonoecy but it is rare compared with andromonoecy. Trimonoecy occurs when hermaphroditic, male and female flowers are present, and is extremely rare. In a similar manner, dioecious populations can be androdioecious or gynodioecious.

Fig. 7.1: (*upper*) *Diplopeltis huegelii* (Sapindaceae), a monoecious species (i.e both female and male flowers on the same plant). *(left)* Female flower, with vestigial stamens that fail to mature, with ovary. *(right)* Male flower with no style or ovary but prominent stamens. To the left is a fruit with three carpels that developed from the female. Flowers 1 cm wide. (*lower*) female inflorescences of dioecious *Allocasuarina* (Casuarinaceae) (*left*) *A. humilis* with 5 mm wide inflorescences. (*right*) *A. campestris* with 4-cm long cone with woody bracts protecting the fruits to the left of the female inflorescence. *Allocasuarina* female flowers have long protruding styles that enhance their ability to collect wind-dispersed pollen produced by male flowers on separate plants.

7.1 Outbreeding Strategies

Fig. 7.2 outlines various breeding strategies that occur within the SouthWest flora. For outbreeding to be effective, pollen needs to be sourced from another flower (allogamy) – either from the same plant (geitonogamy) or different plants (xenogamy). Geitonogamy only works as an outbreeding device when there are genetic differences within the plant due to somatic mutations that are most likely to occur in plants that live for hundreds of years.

To maximise the chance of outcrossing, release of pollen is out of phase with maturation of the stigma (dichogamy) in many hermaphroditic species, and they have evolved a number of strategies to achieve this. Protandry occurs when pollen is shed prior to the stigma becoming receptive and is by far the most common strategy in the Australian flora. Protogyny occurs when the stigma is receptive prior to pollen shedding.

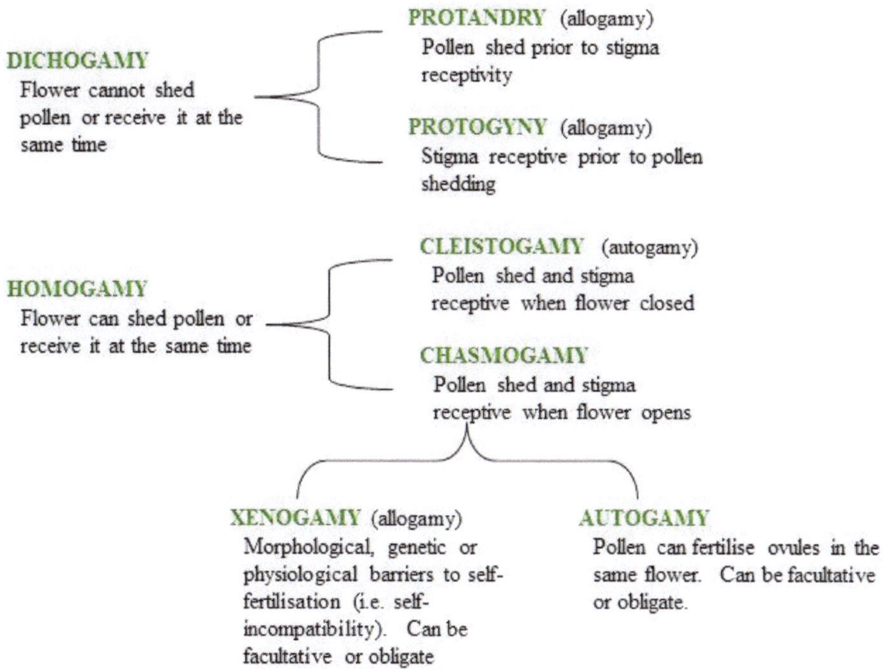

Fig. 7.2: Overview of angiosperm breeding mechanisms.

7.1.1 Protandry

Most protandrous families have secondary pollen presentation strategies that involve pollen deposition onto a pollen-presenting organ (Howell, Slater & Knox, 1993) that is exposed to pollinators prior to a receptive stigma developing. To achieve this the anthers and the pollen presenting floral component must be in close proximity to each other prior to anthesis.

Howell *et al.* (1993) recognized nine secondary presentation strategies that included pollen presenters on the apices of styles with concealed (closed or sunken) stigmas (e.g. Proteaceae), sunken stigmas with pollen placed passively (e.g. Polygalaceae), active pollen-placement (e.g. Asteraceae) or indusial (cup-shaped) stigmatic presenters (e.g. Goodeniaceae). Details of how protandry occurs in selected taxa are now considered.

7.1.1.1 *Eucalyptus* (Myrtaceae)

Flowers have numerous stamens arranged in several whorls on the floral cup (hypanthium). In some species the outermost stamens are replaced by staminodes (sterile modified stamens). When the flower opens, the stamens are mature (Fig. 7.3). Once the pollen has been released the spent stamens move outwards; the central stigma then becomes receptive.

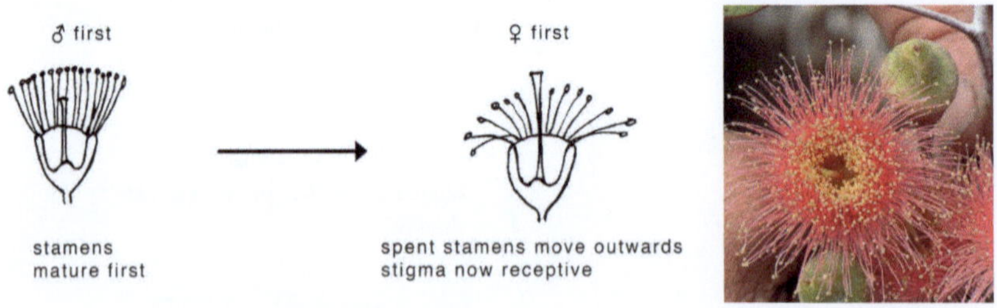

Fig. 7.3: Protandry in the genus *Eucalyptus*. Photograph is of a recently opened flower of *E. caseia*, 3 cm wide, with inner stamens erect.

7.1.1.2 *Pimelea* (Thymelaeaceae)

In the open flower the pollen-bearing anthers are held erect above the floral tube (Fig. 7.4), releasing their pollen as a pollinator brushes past the anthers. Over time the spent anthers fall away and the style elongates bearing a receptive stigma.

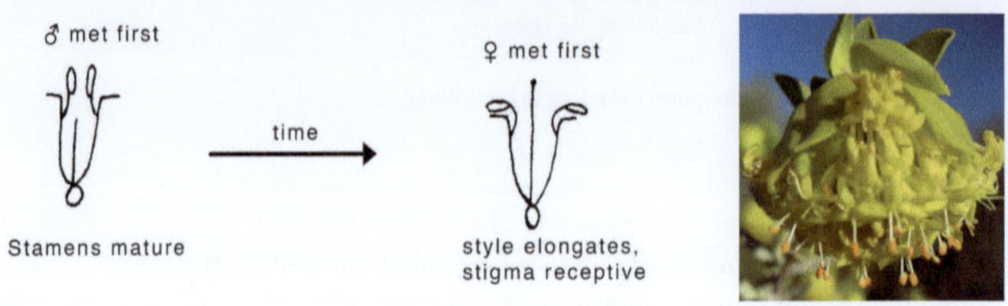

Fig. 7.4: Protandry in the genus *Pimelea*. Photograph is of *Pimelea aeruginosa*, capitulum 3 cm wide.

7.1.1.3 Goodeniaceae

The family Goodeniaceae is characterized by an indusium, a pollen cup surrounding the stigma (Fig. 7.5). Pollen is shed into this cup that then closes. Anthers wither and the style, bearing the indusium, elongates above the anthers. The stigma now develops, pushing the pollen onto the lips of the indusium. The stigma now spreads, pushing away any remaining pollen, and is pollen receptive.

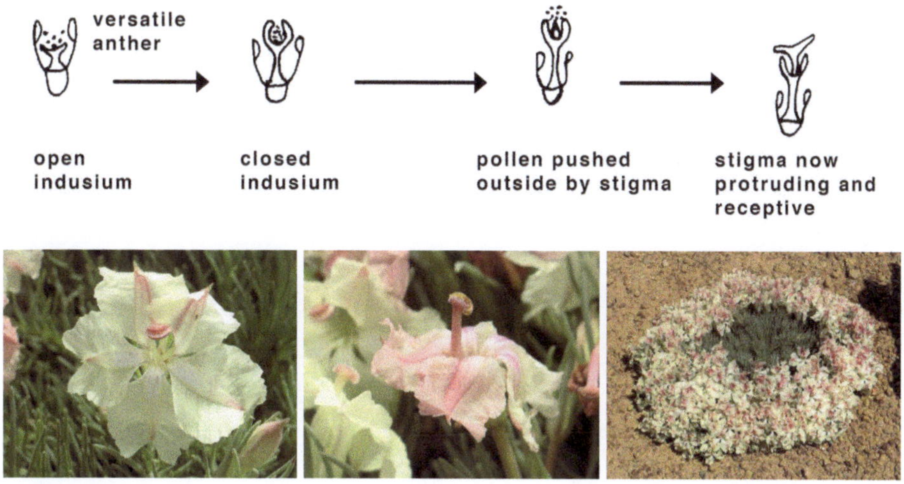

Fig. 7.5: Protandry in the family Goodeniaceae, showing the function of the indusium. Photographs are of *Lechenaultia macrantha* (wreath lechenaultia). (*left*) flower with reddish indusium visible enveloping the hidden stigma, (*centre*) close up of pollen-laden indusium, (*right*) entire plant in full flower. Flowers are 2.5 cm diameter, plant 40 cm diameter.

7.1.1.4 *Banksia* (Proteaceae)

The stigma is shielded by a stigmatic groove (or cleft) that is part of the pollen presenter (Fig. 7.6). Pollen is initially deposited onto the presenter at anthesis (as in *Grevillea* below). Later on, the groove opens revealing the receptive stigma, pushing away any pollen still present around the stigmatic area (Sedgley *et al.*, 1993). There is significant variation among species whether the pollen is deposited over the stigmatic groove in bud or the slit is free of pollen at anthesis (Ladd *et al.*, 1996).

7.1.1.5 *Petrophile* and *Isopogon* (Proteaceae)

Protandry in both genera occurs by the anthers releasing pollen onto the terminal portion of a distinctively enlarged pollen presenter (Fig. 7.7) atop an elongated style. The terminal portion may be brush-like (e.g. *Petrophile linearis*; Fig. 7.7) or glabrous (e.g. *Isopogon dubius;* Fig. 7.7) above which is a stigmatic bulge (Matthews *et al.*, 1999). The swollen basal portion of the presenter may be inversely conical (turbinate) or semi-elliptical. A receptive stigma is formed as the pollen presenter changes colour from yellow to orange-red. This colour change is a signal to pollinators (e.g. native bees) indicating the presence or absence of a reward (pollen) (Lamont, 1985a).

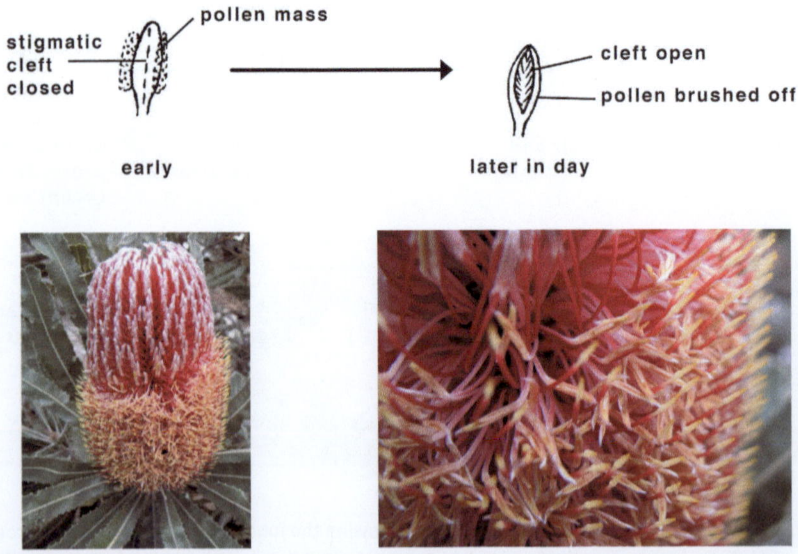

Fig. 7.6: Protandry in the genus *Banksia* involves a pollen presenter, a swollen spindle-shaped tip on their elongated styles. Photographs are of *B. menziesii* (*left*) with open florets a predominantly yellow colour due to the yellow pollen presenters. Unopen florets are predominantly red due to the colour of the style and perianth. (*right*) close up of florets at the transition zone (between open and unopened florets) with the yellow pollen presenter (pollen already removed by birds) prominent on the open florets.

Fig. 7.7: Pollen presenters in (*left*) *Petrophile linearis* and (*right*) *Isopogon dubius*. A change in colour, from yellow to orange-red, occurs after the removal of the deposited pollen, and as the stigma becomes receptive. Note pollen grain adhering to tip of pollen presenter in red condition of *I. dubius*. Pollen presenters 5 mm in length.

7.1.1.6 *Grevillea* and *Hakea* (Proteaceae)

Flowers have four stamens attached to concavities near the apex of the four segments of the perianth (= floral tube). In the closed flower the tip of the style is nestled among the anthers (Fig. 7.8). The style apex is a swollen and flattened pollen presenter.

The presenter surrounds the stigma that is initially submerged and non-receptive. Immediately before the style frees itself from the perianth, pollen is deposited onto the presenter. The pollen is now carried aloft to be presented to the appropriate animal vector. The pollinator brushes the pollen away while seeking nectar at the base of the flower. The receptive stigma emerges a day or two after pollen has been released.

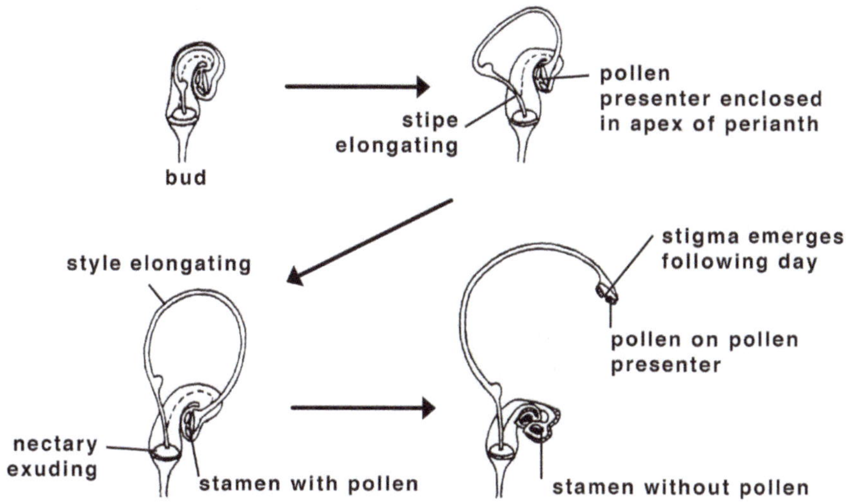

Fig. 7.8: Protandry in the genus *Grevillea*. Diagram depicts the positioning of the pollen presenter in close proximity to the anthers in the closed flower. Immediately before the flower opens, pollen is deposited onto the pollen presenter. The stigma then springs aloft carrying the pollen to a position where it will be scraped off the presenter by a pollinator feeding from the nectary. From Lamont (1982) for *G. leucopteris* that is mainly pollinated by night-flying beetles (*Pachytricha* spp.) that spring open the flowers as they move over the panicle and deposit pollen on flowers in the pistillate condition. Photograph shows various stages of flowering in *G. wilsonii*, fully open flower 4 cm long.

7.1.1.7 *Persoonia* (Proteaceae)

Unlike *Grevillea* and *Banksia* (Proteaceae), *Persoonia* lacks a pollen presenter. Protandry in *Persoonia* occurs as a result of the empty anthers (fused to the perianth) folding away, to expose the receptive stigma (Fig. 7.9).

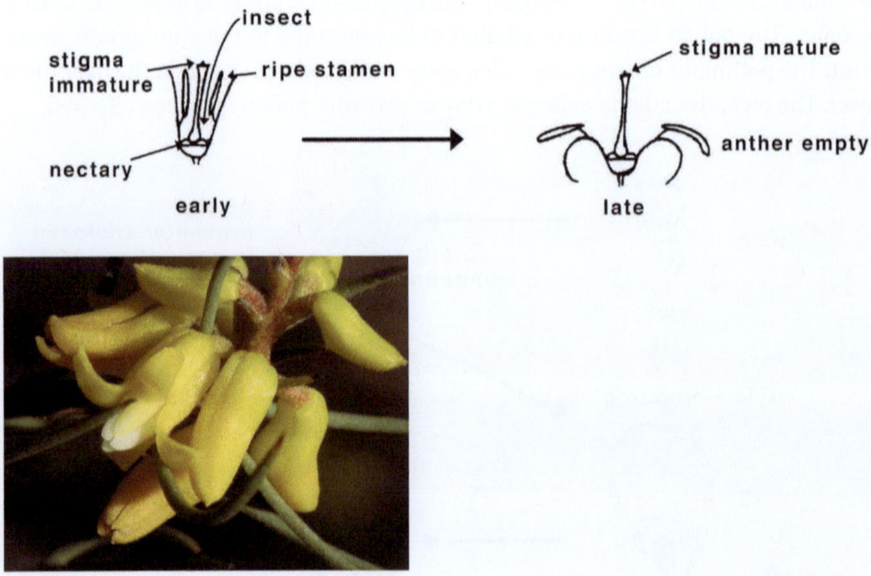

Fig. 7.9: Protandry in the genus *Persoonia*. Flowers, 15 mm long, are of *Persoonia saundersiana*.

7.1.1.8 *Stylidium* (Stylidiaceae)

Stylidium has its anthers and style fused into a recurved column (called the androgynoecium). Nectar-seeking insects press against the flanges at the base of the column and trigger it. The column emerges from between the anterior petals, bending 270° in 10–30 milliseconds, then slowly resets to its original position taking from 200 to 700 s (Findlay & Findlay, 1975; Fig. 7.10). This movement is produced by changes in the curved region (bend) of the column, caused by osmotically driven changes in cell volume resulting from a net influx of K^+ ions from the surrounding tissue (Findlay & Findlay, 1984). A net efflux conversely makes the cell shrink, causing the column to reset. The bend region is also strikingly different anatomically from the rest of the column that caters for the strengthening and movement of the column (Findlay & Findlay, 1975; 1989).

In the photograph of *Stylidium scandens* (flower 1 cm wide) (Fig. 7.10) the flower on the left has the column ready to be triggered, so it is hidden behind the petals. The column has been triggered to fire by a visiting fly in the flower on the right. Pollen is shed onto the insect by force. When the stigma emerges, pushing aside the empty anthers, the column picks up pollen from a visiting insect. The location of pollen deposition on the insect's body varies between *Stylidium* species, a function of column positioning and reach that creates a pollination niche between species (Armbruster, Edwards & Debvec, 1994).

Fig. 7.10: *(upper)* A Bombyliid fly (*Aleucosia tridentata*) feeding on the nectar of *Stylidium scandens* with the androgynoecious column hidden below the 7 mm long petals (*left*). When triggered by touch the column bends in less than 30 milliseconds, hitting the fly with considerable force. In the left image pollen is visible on the fly's dorsal abdomen confirming it has previously visited another *Stylidium* recently, although the positioning of the triggered column (*right*) on the fly suggests it may have been a different species. Images provided by Lochman Transparencies. (*lower*) Recurved column of *Stylidium bulbiferum* (arrows) before and after it has been triggered by touch. Flower 15 mm across.

7.1.1.9 *Eremophila* (Scrophulariaceae)

Pollen is presented to the bird or insect when they first visit the open flower and later the stigma emerges so that the pollinator brushes against this first, donating any pollen that it is carrying to the receptive flower (Fig. 7.11).

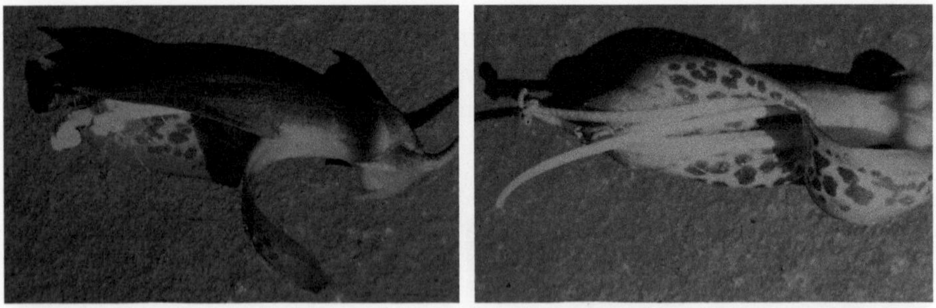

Fig. 7.11: Bird-pollinated *Eremophila maculata* (latin for spotted in reference to the markings on the inside of the corolla) flower, 4 cm long, in male condition, with ripe anthers brushing against the bird when it probes for copious nectar in the 'gullet' (*left*) followed after a day by the flower in the female condition with the stigma now protruding beyond the spent anthers so this is met first when the bird visits (*right*).

7.1.2 Protogyny

Protogyny, in combination with self-incompatibility, is a feature of *Anthocercis gracilis* (Solanaceae), a threatened species. Protogyny prevents autopollination, whereas self-compatibility avoids geitonogamous self-fertilisation (Stace, 1995). This species can be classified as obligately xenogamous as it must receive pollen from another plant for successful seed set.

7.2 Autogamy

Fringe lilies, *Thysanotus* (Fig. 7.12) and other members of the Asparagaceae, press their perianth segments onto the subtended stamens, pushing the anthers onto the stigma later in the day to ensure pollination, if it has not already occurred *via* their bee pollinators. This is secondary to outbreeding if it has not already occurred and at least ensures that fertilisation occurs even in the absence of pollinators (default selfing).

Pollination syndromes in *Thelymitra* species (Orchidaceae) (Fig. 7.12) vary from obligate autogamy to xenogamous insect-mediated pollination. Some species do not open if there is inadequate sunlight (hence the common name sun orchids), opening in response to a change in temperature, and may close over later in the day. When this happens, the flower can self-pollinate, as the edge of the stigma recurves onto the anther, collecting its own pollen (another example of spontaneous autogamy). This is facilitated in *Thelymitra* as, unlike most orchids, its pollen grains do not cohere into a pollinium but remain loose and easily spread. Some cleistogamous species produce flowers that do not open at all later in the growing season and self-pollinate, such as *Pavonia hastata* (Malvaceae).

Fig. 7.12: Flowers of facultative autogamous species. (*left*) *Thysanotus multiflorus*. Flower is 1.5 cm long. (*middle*) Various permutations of autogamy prevail in *Thelymitra*. Pictured *is T. villosa*. Flower is 3 cm long. (*right*) *Calochilus uliginosus* is a terrestrial orchid possessing a labellum covered in dense, long red or blue hairs. All *Calochilus* species are Scollid wasp-pollinated, but will self-pollinate if wasp-mediated pollination does not occur within the day. Flower is 1.5 cm long. *Thelymitra* and *Calochilus* provided by Allen Lowrie.

7.3 Attracting Pollinators

The renowned spectacular flowering and wide range of colours in the SouthWest flora is the outcome of the concerted efforts of plants to attract pollinators. The flora is characterised by many small flowers crowded into simple or compound inflorescences, e.g. spikes, racemes, axillary clusters, umbels, cymes, panicles and corymbs.

Large solitary flowers, e.g. *Eucalyptus macrocarpa* and the related *E. rhodantha* (Fig. 7.14), with flowers up to 7 cm diameter, are extremely rare. Some inflorescences consist of compact groups of flowers surrounded by colourful or conspicuous imbricate (overlapping) bracts so that each head looks like a large flower, called a pseudanthium. The inflorescence of *Pimelea physodes* (Thymelaeaceae) is surrounded by an involucre of large red/pink bracts (Fig. 7.13) that attract nectar-feeding birds. It has been suggested (Keighery, 1975) that the floral bracts mimic those of co-occurring bird-pollinated *Darwinia* (Myrtaceae), as most pimeleas in the SouthWest are butterfly pollinated.

Diplolaena (Fig. 7.13) and *Chorilaena* are the only members of the Rutaceae with congested flowers surrounded by bracts. In *Diplolaena* the inflorescence remains open for 7–10 days, and the long stamens with orange-red anthers contrast with the hairy green bracts. The flowers are well supplied with nectar and attract small beetles, ants and native bees, as well as honeyeater birds. Flowers of *Darwinia* (Myrtaceae) often are surrounded by colourful red/yellow/white bracts (Fig. 7.13). The inflorescences are often pendent, with honeyeaters alighting on the stout supporting stem to probe for nectar.

Although not always as prominent as shown here, flower heads of *Dryandra* (Proteaceae) (now considered a subgenus of *Banksia*) are surrounded by floral bracts (Fig. 7.13). These not only take the place of individual perianths in attracting pollinators, but also act as landing platforms for pollinators and create a pool of nectar within the cup—especially as a reward for large vertebrate pollinators (birds and marsupials).

Other species with petal-like bracts include *Actinodium cunninghamianum* (Myrtaeae), *Andersonia echinocephala* (Ericaceae), and the genus *Johnsonia* (Hemerocallidaceae). In addition, some flower clusters are subtended by a single large leaf produced as part of normal vegetative growth, but because of their bright colours (*e.g. Hakea victoria*) or shell-shape (*e.g. Hakea conchifolia, H. cucullata*) (see Chapter 8) may have similar functional properties as bracts.

Fig. 7.13: Large floral bracts aid pollination. (*left* to *right*) Red-coloured bracts attract birds as in *Pimelea physodes* (Thymelaeaceae) (flower head 5 cm long), *Diplolaena grandiflora* (Rutaceae) (flower head 4 cm wide), and *Darwinia meeboldii* (Myrtaceae) (flower head 5 cm long) are both bird-pollinated. *Banksia* (*Dryandra*) *proteoides* (*far right*) produces orange-brown floral bracts surrounding a head of florets (capitulum is 10 cm long) that is pollinated by small mammals (marsupials and rodents).

7.4 Flower Colour Change

Flower colour change is a mechanism that maximises the efficiency of pollination by maintaining the overall attractiveness of the plant to pollinators from a distance (instead of the spent flower parts turning brown or abscising) but visually cueing pollinators to only visit the rewarding and sexually active flowers once arriving at the inflorescence. The timing of colour change corresponds to a loss of stigma receptivity and nectar production, completion of pollination and onset of seed formation. It is widespread in the SouthWest flora, particularly in the Myrtaceae and Proteaceae (Lamont, 1985a; Weiss & Lamont, 1997).

It can take from 2 to 30 days to complete the colour change depending on the species. For *Conostylis setosa* (Haemodoraceae) individual flowers change from white to dark red (Fig. 7.14), where the red coloration is a signal for the visiting insects to avoid non-rewarding flowers. For *Hakea aculeata* (Proteaceae) the outer florets are the first to change colour (from white to yellow) (Fig. 7.14), enticing the insect pollinator towards the younger nectar-producing, and receptive, flowers in the centre of the inflorescence. The phenomenon is also expressed among some bird-pollinated species, the most notable of which is *Banksia ilicifolia* where the entire inflorescence changes from yellow (bearing nectar and pollen) to deep red (no nectar and pollen already removed) over 4 to 7 days (Fig. 7.15). Heads stay red for 10 days and then begin to senesce, giving the chance for nutrients to be recycled to the developing fruits, though seed set is only 1% of florets. In one study, 27% of flower heads in the stand were in the yellow condition but they received 94% of pollinator visits whereas 64% of flower heads were red but received no visits (Lamont & Collins, 1988). This species is xenogamous and in the absence of honeyeaters, seed set is negligible.

A complete list of species that display flower colour change is provided in Weiss & Lamont (1997).

Fig. 7.14: (*upper left*) *Conostylis setosa* (Haemodoraceae). Head is 4 cm wide. (*upper right*) *Melaleuca radula* (Myrtaceae). Flowers are pink ageing to white. Spike is 5 cm long. (*lower left*) *Verticordia huegelii* (Myrtaceae). Flowers are creamish-white ageing to a dark maroon. Note the flowers form an overall attractive surface (corymb) of colour. Individual flowers are 12 mm wide. (*lower right*) *Hakea aculeata* (Proteaceae) flowers are white aging to yellow. Axillary cluster is 5 cm wide.

Fig. 7.15: *Banksia ilicifolia* (Proteaceae), one of the few species that exhibits flower colour change at the inflorescence rather than individual flower level and, in addition, is predominantly visited by birds. (*left*) young flower head, 5 cm wide, and (*right*) 5-day older flower head. That it is the only one of five co-occurring banksias that shows colour change has been attributed to the facts that it has by far the smallest flower heads and its distribution is much patchier than the others (restricted to winter-damp depressions). By directing birds to the most rewarding heads, colour change appears to enhance foraging and pollination efficiency without reducing the tree's overall attractiveness since yellow and red heads are equally visible, so increasing its relative fitness.

7.5 What Role Do Extrafloral Nectaries Play?

Extrafloral nectaries (EFNs) are sugar- and amino-acid-secreting nectaries that are located outside the flower. Unlike tropical floras, EFNs are not common in the SouthWest. Since there can be a strong correlation between species that are ant-dispersed and the presence of EFNs (Mimosoideae, *Adenanthos*, Euphorbiaceae), it is surprising that EFNS are not more widespread since the SouthWest is recognized as the world centre of diversity for species with elaiosomes (Chapter 10, Berg, 1975). While elaiosomes are a minimal water and nutrient drain on the plant, EFNS are potentially the reverse, so this might explain their abundance in better resourced floras but restricted occurrence here. Nevertheless, they are present in a wide range of taxa (Table 7.1) and are probably universal in the huge genus, *Acacia*.

Attracting herbivore-deterring ants is usually considered the main function of EFNs. The EFNs on *Adenanthos cygnorum* do attract many ant species (Fig. 7.16) but also the parasitoid wasp, *Campoletis* sp. (Ichneumonidae). Both insect groups deter xyloryctid moths from consuming new foliage and stem apices: ants no doubt devour the eggs that can sometimes be seen affixed to the leaves while the larvae are favoured hosts for the wasp (Lamont, 1989; Chapter 10.).

Table 7.1: List of selected species that possess extrafloral nectaries in the SouthWest. Rachis refers to the midrib that supports the pinnae that arise in pairs among compound leaves, usually with an extrafloral nectary at the axil of each pair.

Species	Family	Type	Location on plant	Visitors	Reference
Acacia aneura	Fabacaeae, Mimosoideae	sunken	phyllode	ants	Lamont, 1979
Acacia lasiocalyx	Fabacaeae, Mimosoideae	sunken	phyllode	?	Lamont, 1979
Acacia pentadenia	Fabacaeae, Mimosoideae	sessile	compound leaf, rachis	?	Lamont, 1979
Acacia iteaphylla	Fabacaeae, Mimosoideae	sessile	phyllode	crusader bugs (exotic)	Lamont, 1979
Acacia myrtifolia	Fabacaeae, Mimosoideae	sessile	phyllode	ants	Marginson, Sedgley, Douglas, *et al.*, 1985
Acacia saligna	Fabacaeae, Mimosoideae	sessile	phyllode	ants	Lamont, 1979
Acacia pulchella	Fabacaeae, Mimosoideae	stalked	leaf axil, tip of rachis	?	Lamont, 1979
Adenanthos cygnorum	Proteaceae	sessile	tips of leaf segments	ants, wasps	Lamont. 1989
Adenanthos cuneatus	Proteaceae	sessile	tips of leaf teeth	ants	This book
Paraserianthes lophantha	Fabacaeae, Mimosoideae	sessile	compound leaf, rachis	ants	Lamont, 1979
Senna artemisioides subsp. *filifolia*	Fabaceae, Caesalpinioideae	sessile	compound leaf, rachis	?	Lamont, 1979
Hardenbergia comptoniana	Fabaceae, Faboideae	sessile	compound leaf, stipules, stipels	ants	Lamont, 1979
Stylidium breviscapum	Stylidiaceae	stalked	simple leaf, tip	?	Lamont, 1979
Pteridium esculentum	Dennstaedtiaceae	sessile	frond, nodes of lower pinnae	ants	Lamont, 1979
Hemigenia saligna	Lamiaceae	sunken	Axil between pair of leaves	ants	Lamont, 1984b
Chamelaucium uncinatum	Myrtaceae, Leptospermoideae	sunken	Axil between pair of leaves	ants, bees	Obrien, 1995

Fig. 7.16: Extrafloral nectaries (EFNs) are universal on the leaves and phyllodes of SouthWest acacias. They can be grouped into three types: (*left*) sessile (e.g. 2-mm-long in *Acacia iteaphylla*), (*middle*) sunken (e.g. 1.5-mm-wide in *A. aneura* which also possesses T-shaped and glandular hairs in lines along its terete phyllode) and (*right*) stalked (e.g. 2.5-mm-long in *A. pulchella* also showing thorn in axil). Ants feeding from EFNs: (*left*) *Iridomyrmex bicknelli* on *Adenanthos cygnorum* – note how the 2-mm-wide EFNs surround the new foliage in the centre, and (*right*) *Iridomyrmex* aff. *conifer* on *A. cuneata* – note flower at lower left surrounded by young 30-mm-long, red leaves perhaps serving as a supplementary attractant to bird pollinators? Nectar produced by ant-attracting EFNs is not accessible to birds as the amount is so small.

Ants may be especially important at the seedling stage as nectar exudation is particularly prominent then (personal observations) but relevant research is lacking. However, the synchrony between exudation and vegetative growth is lost in such species as *Acacia aneura*, Australia's most widespread acacia (Lamont, 1979). Here, exudation is greatest in summer when this species is flowering and ants are most active on the plant then. It is possible that insect florivores may be attracted to the plant at that time but there is evidence for other acacias that they have a role in pollination. In a study of southeastern Australian acacias, 24 wasp taxa (Sphecidae and Tiphiidae) were recorded as foraging at the EFNs before visiting the nectarless inflorescences whereas the far-more-common, 27 bee taxa (Halictidae and Colletidae) ignored the EFNs and only foraged for pollen (Bernhardt, 1987). Studies on Australia's floral emblem, *Acacia pycnantha*, and other species with compound inflorescences, shows that the EFNs attract honeyeaters, silvereyes and thornbills at that time and acacia pollen has been identified among their feathers (Knox *et al.*, 1985; Vanstone & Paton, 1988). Since nectar production by the flowers themselves is negligible this is a novel way to attract such highly efficient pollinators to the plant.

Acacia saligna, the most widespread acacia in the SouthWest, may harbour 69 herbivorous insect species but the EFNs also attract 19 species of aggressive ants (J.D. Majer, personal communication). When experimental plants were damaged to simulate herbivory, they began exuding nectar from their EFNs. *Iridomyrmex chasei* ants given access to these damaged plants were 10 times more likely to survive than those given access only to undamaged plants and survived as well as those provided with sugar solution only (Lach *et al.*, 2009). The induction of EFN exudation in response to herbivory and the subsequent attraction and sustenance of aggressive ants indicate a powerful strategy for minimizing the level of herbivory on this species, a mutualism that relies on the reliable presence of sugar-seeking ants.

8 Pollination Syndromes

The vast majority of species in the SouthWest flora requires animal visitors to their flowers to transfer pollen and promote outbreeding (known as zoophily) (Phillips *et al.*, 2010). However, major monocot families – the grasses (Poaceae), sedges (Cyperaceae) and wirerushes (Restionaceae) have abundant pollen that is solely wind dispersed (anemophily) and also have feathery stigmas that catch pollen. Anemophily also occurs in the dicot familes of Sapindaceae (*Dodonaea*), Surianaceae (*Stylobasium*) and the fire ephemeral, Gyrostemonaceae (Keighery, 1981). For zoophilous species, flower colour, shape and size, and the amount of nectar produced are important determinants of the type of animal they attract, and in some cases they only utilize a single pollinator species (e.g. sexually deceptive orchids) (Table 8.1).

Table 8.1: Typical floral syndromes of the SouthWest flora that are adapted for pollination by insect, bird or mammal species.

	Insect	**Bird**	**Mammal**

Example	*Hypocalymma angustifolium*	*Banksia prionotes*	*Banksia nutans*
Anthesis	Diurnal, nocturnal	Diurnal	Nocturnal
Colour	Blue, white, yellow,	Red, yellow, green	Purple, brown, orange
Intensity	Bright	Bright	Dull
Position	Prominent	Prominent	Hidden
Inflorescence	Soft	Firm, wiry	Firm, wiry
Inflorescence size	Small	Large	Large
Nectar - volume	Small	Large	Large
- sugar content	Low	High	High
Odour	Strong (sweet)	Scentless	Strong (musky)

The variety of flower colours in the flora indicates appeal to a wide range of possible vectors. Insects (entomophily) are most sensitive to the ultraviolet and blue end of the light spectrum. Birds (ornithophily) are most sensitive to the red end and see in a similar range to humans.

For example, 75% of *Eremophila* species (Scrophulariaceae) are entomophilous with a protruding lower lip as a landing platform and are coloured blue, purple or white, while the remainder are ornithophilous with the lower lip depressed away from the throat with exserted stamens and style and are red, orange, yellow or green (Chinook, 2007). Mammals see in a similar range to humans but rely much more on their strong sense of smell, unlike birds. Thus, three basic pollination syndromes may be recognized in the SouthWest flora (Table 8.1).

8.1 Insect-pollinated Species

The range of insect pollination strategies occurring within the SouthWest flora is provided in Table 8.2., with examples of insect-pollinated flowers in Fig. 8.1. The main insect pollinators of the SouthWest flora are hymenopteran (bees, wasps), dipteran (particularly bee flies, but also other flies, mosquitos, gnats), coleopteran (beetles, particularly scarab beetles; Fig. 8.2) and lepidopteran (butterflies, moths) species. In this book we highlight the possible coevolutionary responses of plants to native bees and wasps.

8.1.1 Case Study: *Grevillea leucopteris*

Grevillea leucopteris (Proteaceae) has all the characteristics of a species pollinated by night-active insects (Fig. 8.2) including emitting a musky odour and the production of copious nectar at dusk that peaks by midnight (Lamont, 1982c). Giant scarab beetles (*Pachytricha* spp.: Melolonthidae) are lured to the plants from great distances by the smell and then guided by the white flowers that stand out against the dark sky. As they probe for nectar they spring open the flowers, whose turgidity greatly increases at nightfall, and pollen brushes against their thorax and wing covers. Many moths are also attracted but they are ineffective as pollinators, while crickets and honey possums are effective but poorly mobile by comparison. *G. leucopteris* is self-compatible and spontaneous autogamy and geitonogamy may still result in seed set if xenogamy fails.

Table 8.2: Summary of floral characteristics of entomophilous pollination subcategories (adapted from Armstrong, 1979)

Syndrome	Colour range	Structure	Access to reward	Odour	Species examples
Canthorophily (beetle pollinated)	Dull coloured; white-cream	Flowers facing upward, or landing platform provided	Easy. Reward usually pollen rather than nectar	Strong – fruity/ spicy	*Cyanicula gemmata* (Orchidaceae) *Hibbertia hypericoides* (Dilleniaceae) *Philotheca* spp. (Rutaceae)
Myophily (flies or gnat pollinated)	Various colours; pale or dull	Depends on pollinator mouthpart length	Easy	Nil or strong – foul smelling	*Pterostylis* spp. (Orchidaceae) *Monotoca tamariscina* (Ericaceae) *Leucopogon verticillatus* (Ericaceae) *Hakea denticulata* (Proteaceae) *Grevillea gordoniana* (Proteaceae)
Melittophily (bee pollinated)	White, cream, yellow, blue, purple	Deep and robust; held terminally on flowering stems in clusters or singularly in axils	Difficult, often concealed	Nil or slightly sweet	*Verticordia nitens*, *V. aurea* (Myrtaceae) *Conostephium pendulum* (Ericaceae) Fabaceae (many spp. in the three subfamilies) *Conospermum* spp. (Proteaceae)
Myrmecophily (ant pollinated)	Various colours; Pale or dull	Small, clustered, usually close to the ground	Easy or no reward	Nil	*Microtis parviflora* (Orchidaceae)
Sphecophily (wasp pollinated)	Dull, non-descript	Single blooms terminal on flowering stems. Structure unique to specific pollinator	No reward; or if present easily accessible	Nil (may emit pheromones to attract a male wasp)	*Drakaea* spp. (Orchidaceae) *Cryptostylis* spp. (Orchidaceae) *Darwinia diosmoides*, *D. pauciflora* (Myrtaceae) *Hemiandra pungens* (Lamiaceae)
Phalaenophily (moth pollinated)	Pale-white, cream, or dull red	Pendulous or held at angle	Difficult. Long distance to nectar source	Strong, sweet	*Lysinema ciliatum*, *L. elegans* (Ericaceae)
Psychophily (butterfly pollinated)	Pink, magenta-red. Vivid colouring	Terminal and erect. Medium sized single flowers or clustered small flowers	Difficult. Long distance to nectar source	Nil, or subtle	*Calytrix* spp. (Myrtaceae) *Pimelea* spp (Thymelaeaceae)

Grevillea gordoniana (Proteaceae) with blowfly

Scaevola species (Goodeniaceae)

Acacia pulchella (Fabaceae)

Hibbertia hypericoides (Dilleniaceae)

Rhodanthe chlorocephala (Asteraceae)

Melaleuca robusta (Myrtaceae)

Verticordia acerosa (Myrtaceae)

Banksia lemanniana (Proteaceae) with *Cryptocheilus* wasps

Viminaria juncea (Fabaceae)

Fig. 8.1: Examples of the taxonomic, stuctural and colour diversity within insect-pollinated flowers in the SouthWest flora.

Fig. 8.2: *Grevillea leucopteris* has all the characteristics of a species pollinated by night-active insects. (*upper left*) two plants with naked scapes holding thousands of white flowers during spring up to 4 m above the scrub-heath in the northern sandplains. (*upper right*) Flowers in terminal panicles of racemes, 8-12 cm long, open towards the apex suggestive of an adaptation to flying pollinators. (*lower*) Giant scarab beetles, *Pachytricha* spp., 4 cm long, are the main pollinators.

8.1.2 Bee Pollination

Hymenopteran species, primarily bees, tend to visit only one flower type to obtain pollen and/or nectar, using scent and visual cues (Chittka *et al.*, 1999). Floral colour properties, revealed by their spectral reflectance, have a close affinity with the ability of bees to visualize and discriminate colour patterns (Dryer *et al.*, 2012). Because this synergy exists in both the Northern Hemisphere and Australian floras independently, it is hypothesized that bee visual adaptability has been a major evolutionary driving force for the rapid speciation and radiation of Australian, insect-pollinated, angiosperm taxa (Dryer *et al.*, 2012). Australia has a high number of species in the bee family Colletidae (Batley & Hogenboorn, 2009) an ancient group of bees with Gondwanan origins that evolved during the Cretaceous (72–82 Ma) with an accelerated period of radiation worldwide commencing during the Oligocene and Neogene (Almeida *et al.*, 2012). The Australian endemic subfamily Euryglossinae and the Australian biodiverse subfamily Hylaeinae both evolved 48–54 Ma. Bee pollination (melittophily) is considered an ancestral trait in the tribes Mirbeliae and Bossiaeeae (Fabaceae) (Toon *et al.*, 2014), Haemodoraceae (Hopper *et al.*, 2009), Australian Ericaceae (Johnson, 2012), Myrtaceae (Lughadha & Proenca, 1996) and several other families.

Many native SouthWest bees adapt either morphologically or behaviourly to their food sources, or at least trade-offs between bee and floral characteristics exist that allow the development of exclusive plant-pollinator relationships. They may visit a taxonomically-diverse range of flowers or specifically visit plants in the same genus (oligolectic) or just a particular species (monolectic). Bees of the Australian endemic *Euryglossa* (Colletidae) are considered important pollinators of the Myrtaceae (Exley, 2004). The oligolectic *Euryglossa tubulifera* accesses nectar from the predominantly bird-attracting *Calothamnus* (Myrtaceae) flowers (see Fig. 8.11) because of its enlarged mouth-feeding parts that join to form an elongated tube (Houston, 1983), whereas the monolectic *E. semaphore* only visits the pink-flowered *Pileanthus filifolius* (Houston, 1992). Twenty *Euryglossa* species seem to have adapted to feeding from blue-violet-flowered *Eremophila* (Scrophulariaceae) species through enlarged labial palps and longer heads (Exley, 1998); thus euryglossin bees are not exclusive pollinators of the Myrtaceae. Bee species from the megachilid genus *Chalicodoma* have been observed visiting the pink flowers of *Calytrix fraseri* (Myrtaceae) (Hawkeswood, 1993), a species that has floral features associated with butterfly pollination (Morcombe (1968) has an image of a butterfly visiting *C. fraseri*).

Verticordia (Myrtaceae) flowers have a special affinity with bees as well as flies and beetles (Hawkeswood, 1993; Houston *et al.*, 1993; Exley, 2004), and occasionally are adapted for honeyeater pollination (e.g. *V. grandis, V. staminosa*). Table 8.2 lists some of the bee species know to visit *Verticordia* flowers. Many of the euryglossin *Dasyhema* are associated with different species of *Verticordia* resulting in the recognition of new species of this bee (Exley, 2004). Some are named after the *Verticordia* they are associated with (e.g. *D. spicata* visits *V. spicata*) (Table 8.3). This does not necessarily imply that these *Dasyhema* are exclusive pollinators of their associated *Verticordia*. For example, although *D. forrestii* has been recorded visiting only *V. forrestii* flowers (Exley, 2004) the jewel beetle, *Castiarina forrestii* (Buprestidae), is also a visitor (Bellamy *et al.*, 2013). Flowers of *V. nitens* and *V. aurea* have hooded anthers with a terminal spout that releases a pollen-oil mixture (Houston *et al.*, 1993; Fig. 8.3). Each species is exclusively pollinated by the euryglossin bees *Euryglossa morrisonii* and *E. aureophila* respectively (Houston, 1992; Houston *et al.*, 1993).

Fig. 8.3 *Verticordia* (Myrtaceae) is an advanced genus with highly modified flowers in a family that stretches back 65 million years. All but two of the known 104 species are endemic to the SouthWest (George & Pieroni, 2002; George & Barrett, 2010). (*lower left*) *V. nitens* is a slender shrub that appears top heavy because of the broad surfaces of brilliant orange flowers (corymbs) it produces in early summer atop a thin main stem, 0.5–1.75 m tall. This species is locally abundant in the understorey of banksia low woodland of the Swan Coastal Plain. The corymbs act like a single large flower (pseudanthium) as a device for enhancing attractiveness to pollinators, serving as a landing platform and facilitating movement from one flower to the next. Unlike many other oligolectic verticordias that present their pollen to visitors (Lamont, 1985a), *V. nitens* has a specialist pollinator (monolectic), a tiny solitary bee, *Euryglossa morrisoni* (Colletidae), that is entirely reliant on this verticordia for its nutrition (Houston *et al.*, 1993). (*upper left*) a female bee squeezing the hood of an anther with its mandibles and lapping up the pollen-essential oil mix from the terminal spout through which it is released. Note globs of pollen on its head and thorax. Flowers, 8 mm wide, with their highly reflective petals and hypanthium make the specific name, *nitens* (latin for shining) most appropriate. (*upper right*) a male bee lapping up nectar from the floor of the hypanthium and receiving oily pollen or brushing pollen onto the central stigma. (*lower right*) Scanning electron micrograph showing individual flower of *V. nitens* with its 10 fertile stamens: h = helmet cover over anther, s = spout through which oily pollen is passed to the bee. The dearth of nectar and the inaccessibility of both nectar and pollen appear to explain the lack of visits by larger pollinators and nectar robbers. In the absence of *E. morrisoni*, seed set is negligible.

Table 8.3: Australian native bees recorded as visiting *Verticordia* species. Data provided where both *Verticordia* and bees were identified to the species level. Information obtained from the Australian Bee Pollination Index Query database *via* the Victorian Museum website, www.museum.vic.gov.au. Accessed 15 August 2014.

Verticordia species	Bee species	Bee Family
V. argentea	*Dasyhesma argentea*	Colletidae
V. aurea	*Euryglossa aureophila*	Colletidae
V. chrysantha	*Paracolletes nigrocinctus*	Colletidae
V. chrysantha	*Stenotritus greavesi*	Stenotritidae
V. chrysantha	*Ctenocolletes smaragdinus*	Stenotritidae
V. densiflora	*Exoneura nigrescens*	Apidae
	Exoneura pictifrons	Apidae
	Hylaeus rufipes	Colletidae
	Paracolletes callander	Colletidae
	Paracolletes nigrocinctus	Colletidae
	Lasioglossum sculpturatum	Halictidae
	Lasioglossum lichatum	Halictidae
	Lipotriches australica	Halictidae
	Lipotriches flavoviridis	Halictidae
	Megachile rugosa	Megachilidae
	Ctenocolletes smaragdinus	Stenotritidae
V. densiflora var. *cespitosa*	*Ctenocolletes albomarginatus*	Stenotritidae
V. dichroma var. *dichroma*	*Euryglossa evansi*	Colletidae
	Dasyhesma simulata	Colletidae
V. dichroma var. *syntoma*	*Dasyhesma galbina*	Colletidae
	Dasyhesma syntoma	Colletidae
V. forrestii	*Dasyhesma forrestii*	Colletidae
V. huegelii	*Paracolletes nigrocinctus*	Colletidae
V. lepidophylla var *quantula*	*Dasyhesma lepidophyllae*	Colletidae
V. muelleriana subsp. *muelleriana*	*Dasyhesma muelleriana*	Colletidae
V. nitens	*Euryglossa morrisonii*	Colletidae
V. oculata	*Dasyhesma areola*	Colletidae
V. pholidophylla	*Amegilla cingulata*	Apidae
V. picta	*Paracolletes nigrocinctus*	Colletidae
	Ctenocolletes albomarginatus	Stenotritidae
	Ctenocolletes smaragdinus	Stenotritidae
V. plumosa	*Thyreus waroonensis*	Apidae
	Paracolletes nigrocinctus	Colletidae
	Leioproctus velutinellus	Colletidae
V. polytricha	*Euryglossa evansi*	Colletidae
V. pritzelii	*Lasioglossum instabilis*	Halictidae
V. serrata var. *serrata*	*Euryglossa lucida*	Colletidae
V. spicata	*Dasyhesma spicata*	Colletidae
V. stenopetala	*Ctenocolletes smaragdinus*	Stenotritidae

Anther morphology and function in *Verticorida* was surveyed by Ladd *et al.*, (1999) who recognized three main anther types, with and without oil-producing glands. Not all oil-producing anthers are hooded like *V. aurea* and *nitens*, and there is some relationship within subgenera between the type of pollen presenter and the presence of anther glands. In a study of *Verticordia* pollen morphology, Thornhill *et al.* (2012) noted that pollen with very small colpi (elongated apertures) are more likely to be associated with anther oil-production. No doubt anther modifications are the result of natural selection for maximizing pollen transfer, and thus there are potentially numerous monolectic melittophilous relationships occuring within *Verticordia* many of which we are currently unaware. Pollen structure is unusually variable within the mostly entomphilous *Conostylis* (Haemodoraceae) (Pierce & Simpson, 2009) and has adaptive implications for evolution of pollinator selection over the 14 million years of *Conostylis* existence (Hopper *et al.*, 2009).

Another colletid genus, *Leioproctus*, is an important bee pollinator of the SouthWest flora. Three species (*L. conospermi*, *L. pappus* and *L. tomentosus*) are cryptic obligate feeders on *Conospermum* (Proteaceae) flowers (Houston, 1989) with mouthpart modifications enabling these species to cope with *Conospermum*'s explosive pollen release. In contrast, the monoletic *L. macmillanii* has an unusually elongated head that enables access to nectar of the floral-tubed *Astroloma xerophyllum* (Ericaceae) (Houston, 1991). *Leioproctus* has also been observed visiting the buzz-pollinated pendulous and nectarless flowers of *Conostephium* (Ericaceae) (Houston & Ladd, 2002). Buzz pollination is unusual in this genus because the anthers are not physically available to visiting bees. Instead the sonified vibrations are transferred to the anthers *via* the tapered purple corolla tube for pollen release (Houston & Ladd, 2002).

Buzz pollination occurs in species that require the rapid vibration of the anthers (known as sonication) to release pollen, and is performed exclusively by native bees. Buzz-pollination is a common feature not only of *Conostephium* but also *Coleanthera* (Ericaceae) (Houston & Ladd, 2002), *Hibbertia* (Dilleniaceae) (Tucker & Bernhardt, 2000), *Dianella* (Hemerocallidaceae) (Duncan, Cunningham & Nicotra, 2004), *Solanum* (Solanaceae), *Thysanotus* (Anthericaceae) (Fig. 7.12) and *Tetratheca* (Elaeocarpaceae).

Hibbertia species (Fig. 8.2) are pollinated by female pollen-collecting bees (Colletidae and Halictidae), scarab beetles (Scarabaeidae) and pollen-eating flies (Syrphidae), with four pollination syndromes recognised in the genus that are based on flower size and symmetry (Tucker & Bernhardt, 2000).

8.1.3 Orchid Pollination

The Orchidaceae is exclusively entomophilous, and many SouthWest orchid species are pollinated by a specific insect species. Four pollination-attracting strategies are recognised in South West orchids, and outlined in detail in Table 8.4 and Fig. 8.4:

These are:
1. Nectar-producing flowers with diverse pollinators (low specifity)
2. Flowers that mimic flowers of other species but are themselves non-rewarding (medium specifity)
3. Winter-flowering species that attract fungus-feeding insects (high specificity)
4. Sexually deceptive—parts of the flower mimic female insects (very high specificity)

Table 8.4: Pollination strategies of southwestern Australian orchids (after Brundrett, 2007).

	Low specificity	**Medium specificity**	**High specificity**	**Very high specificity**
Pollinated by	Diverse insect groups	Insects with similar food requirements	Similar insects in the same functional group	Single insect or several similar species
Mechanisms	Nectar/food reward, often scented	Flowers resemble those of species from other plant families (food deception) but with no reward	Fungus mimicry; entrapment	Sexual deception; orchid mimics a female insect in shape and scent (pheromones)
Insect groups	Beetle, bee, fly, wasp, gnat	Beetle, bee, bee-fly, wasp	Fungus gnat, phorid fly, mosquito	Thynnine wasp, Ichneumonid wasp, ant
Representative genera	*Cyrtostylis*, *Microtis*, *Prasophyllum*, *Pyrorchis*, *Caladenia* (some), *Eriochilus* (some)	*Diuris*, *Caladenia* (~ ⅓ spp.), *Cyanicula*, *Eriochilus*, *Thelymitra*	*Corybas*, *Pterostylis* (most), *Rhizanthella*	*Caladenia* (~⅔ spp.), *Calochilus*, *Cryptostylis*, *Drakaea*, *Leporella*, *Paracaleana*, *Spiculaea*

An additional strategy that requires further evaluation are flowers with labellum projections mimicking stamens or pollen-laden anthers. This may occur in *Lyperanthus*, *Praecoxanthus* (Fig. 8.4) and some species of *Cyanicula*, with the deception being a visual cue for what is perceived as presence of a reward that attracts native bees as they approach the flower.

Fig. 8.4: *Praecoxanthus aphyllus* (*upper left*) emits a fragrant floral scent that predominantly attracts bees and prominent yellow calli on a purple and green labellum that superficially resemble pollen-bearing stamens. Flower is 4 cm wide. *Lyperanthus serratus* (*upper right*) is characterised by dense white calli that also superficially represent stamens. Flower is 5 cm wide. (*lower left*) *Ericksonella saccharata*, once classified as a *Caladenia*. Little is known about the pollination of this monotypic species. Flower is 2 cm wide. (*lower right*) *Eriochilus scabra* is probably food deceptive and attracts native bees. Flower 3 cm wide. *Ericksonella* image provided by *Esperence Wildflowers*. All other images provided by Mark Brundrett.

Some species are obligatory self-pollinators (e.g. *Thelymitra vulgaris, Microtis media*, some *Calochilus* species) with flowers open for a few hours. *M. media* subsp. *densiflora* can produce up to 150 densely-packed flowers on a flowering spike, with open, fertilized and in-bud flowers co-occurring (Hoffman & Brown, 2011).

Within the Orchidaceae, the radiation and evolutionary transitions between pollinator strategies (reward vs deception) and rewards offered (nectar/food reward or no reward) occur independently across evolutionary lineages (Jersáková *et al.*, 2006; Johnson *et al.*, 2013). It is generally recognised that providing no nectar reward is an ancestral condition in orchids (Jersáková *et al.*, 2006) and that evolutionary shifts from deceptive to reward strategies implies adaptive responses to combat pollinator limitation issues relating to pollen dispersal and fecundity. Also the shift from food to sexual deception may be driven by selection for more efficient pollination, without compromising the high levels of gene flow that are characteristic of deceptive species (Scopece *et al.*, 2010). To further understand the evolutionary and ecological significance of pollinator syndromes in SouthWest orchids, more comparative studies on how obligate self-pollination, food reward, food deceptive, and sexual deceptive strategies function within a phylogenetic framework are required.

8.1.3.1 Food Rewards

Orchid taxa that produce nectar as food reward are often scented and brightly coloured to attract frequent insect visitors, thus maximising the opportunities for pollen dispersal and pollination to occur including increased probability of selfing *via* geitonogamy. Flower size restricts the type of insect that can access the sweet nectar, an adaptation to prevent wastage and target beneficial pollinators.

Prasophyllum flowers are highly nectariferous and encourage visits by nectar-seeking beetles, flies, wasps and bees (Bernhardt & Burns-Balogh, 1986a; Elliott & Ladd, 2002) (Fig. 8.5). The related genus *Microtis* attracts small wingless ants, wasps, weevils and small flies (Peakall & Beattie, 1991; Hoffman & Brown, 2011) that visit for a nectar reward, whereas *Cyrtostylis* produces minute flowers that are pollinated by micro-flies. *Prasophyllum, Microstis* and *Cyrtostylis* species may have extremely small flowers (< 3 mm in length; Fig. 8.5) that attract micro-insects, with larger flowers (up to 25 mm long) only occurring within *Prasophyllum*. The monotypic SouthWest endemics *Leptoceras menziesii, Ericksonella saccharata* (Fig. 8.4) and some of the non-endemic genus *Eriochilus* may all fit into the food reward category, and all have white or purple perianth parts that act as attractants to native bees.

Fig. 8.5: (*left*) *Cyrtostylis robusta* produces nectar in a small reservoir at the base of the long, pointed labellum that runs down a narrow central groove and entices microdiptera (Hoffman & Brown, 2011). Flower length 15 mm. (*right*) *Prasophyllum regium* flowers are upside down (labellum faces upwards) and up to 25 mm long (Hoffman & Brown, 2011) and in this image a flower is being visited by a wasp. *Prasophyllum* image provided by Mark Brundrett. *Cyrtostylis* image courtesy of *Esperence Wildflowers.*

8.1.3.2 Pollinator Deception

Pollinator deception (also known as food deception) is considered to be an evolutionarily stable strategy within the Orchidaceae (Jersáková *et al.*, 2006). It is an important driver of floral and species diversity within the Orchidaceae, with food-deceptive orchids speciating at a much slower rate than sexually-deceptive orchids (Cozzolino & Widmer, 2005). Floral mimicry, or floral deception, may be modelled on mimicking one species or a suite co-flowering species, or attracts pollinators with commonly occurring colours, markers and scent (Schiestl, 2005).

Dafni & Bernhardt (1990) placed food deception into three groups. Generalist deception does not mimic a co-occurring model and thus attracts generalist insect foragers that will visit a range of co-flowering species. Floral mimicry occurs where the orchid flower mimics an individual, or a guild of related, taxa and is pollinated by insects that visit it. The third category is where the orchid flowers at the same time as unrelated species that offer nectar and/or pollen rewards to pollinators that the orchid can utilise. Food-deceptive orchids attract a guild of local insect pollinators, taking advantage of potential pollinator availability despite the risk of pollinator competition for pollinator visits. There is potential for co-occurring food deceptive orchids to share pollinators and thus increase the chance of natural hybrids occurring.

Caladenia contains species with two pollination strategies – food deception and sexual deception (Stoutamire, 1983; Phillips *et al.*, 2009). Food deceptive caladenias may be brightly coloured (e.g. *C. flava*, *C. latifolia*, Fig. 8.6) or dull (*C. bicalliata*), are nectarless and have some capacity to self-pollinate, suggesting that this is a strategy to ensure pollination occurs in the absence of visiting pollinators (Phillips *et al.*, 2009). It is proposed that food deceptive caladenias are not mimicking other species (non-model mimicry) as no other nectar rewarding taxa are morphologically similar in the SouthWest (Phillips *et al.*, 2009). Like the rewarding strategy, food deception may be viewed as evolutionary responses to overcome pollen limitations arising from infrequent pollinator visits and to maximise visitations between clonal plants.

The SouthWest endemic and monotypic *Praecoxanthus aphyllus* attracts native bees by a highly fragrant scent (Hoffman & Brown, 2011) and the display of a predominantly purple labellum with yellow calli on a purple and green labellum that partially resemble pollen-bearing stamens (pseudoandroecia) (Fig. 8.4). *Praecoxanthus* is said to be nectarless (M. Brundrett, *pers. comm.)* and probably utilises its pseudoanthers, coloured labellum and white perianth parts to attact visitoring insects (thought to be mainly native bees) to an absent reward. Native bee pollination has been observed in *Pheladenia*, *Eriochilus*, and *Cyanicula* (along with beetles) (Hopper & Brown, 2004) and are potentially floral mimics, although the type of deception utilised has yet to be investigated. The relatively dull-coloured flowers of *Lyperanthus serratus* have been reported to produce nectar (Elliott & Ladd, 2002), but may be utilising their prominent and dense white calli at the apex of their labellum as a form of stamen mimicry to attract native bees (Fig. 8.4).

Fig. 8.6: (two *upper rows*) Flowers of *Caladenia flava* and *C. latifolia* offer food rewards to potential insect pollinators, in contrast to the sexually-deceptive flowers of *C. roei, C. discoidea, C. attingens* ssp. *attingens* and *C. arenicola*. All sexually-deceptive *Caladenia* species are pollinated by thynnine wasps (Phillips *et al.*, 2009). Left image shows male *Phymatothynnus* wasps visiting flowers after being lured by sexually-deceptive volatile semiochemicals (as interpreted by Phillips *et al.*, 2009). (*third row*) *Prasophyllum* attracts insect pollinators with a sweet odour. Pictured is *Prasophyllum gracile*, the smallest member of the genus. Flowers are 5 mm in length. *Diuris longifolia* flowers (*middle*) utilise mimicry of co-occurring legume flowers (e.g. *Daviesia triflora; right*) to attract pollinators. (*last row*) The non-rewarding blue flowers of *Thelymitra crinita* appear to mimic native lilies (e.g. *Dianella*) or native irises (e.g. *Patersonia*). In contrast, the cream-flowered *T. antennifera* attracts pollinators of *Hibbertia* and *Goodenia* (Dafni & Calder, 1987; Bernhardt, 1995) – all reradiate strongly in the UV range. *T. apiculata* (*right*) closely resembles the flowers of *Calectasia* (Dasypogonaceae). Flower is 3 cm wide. Images of *C. discoidea* and *T. apiculata* provided by Mark Brundrett. Image of *C. attingens* provided by Allen Lowrie.

For food deceptive *Thelymitra* species the deceit involves the mimicry of visual (flower colour) and olfactory (sweet odour) cues of co-occurring and co-flowering genera to attract male bees (Bernhardt & Burns-Balogh, 1986b; Dafni & Calder, 1987; Sydes & Calder, 1993). The blue flowers of the SouthWest endemics *T. crinita* (Fig. 8.6) and *T. macrophylla* resemble those of blue-flowered lilies (e.g. *Orthrosanthus laxus, Patersonia occidentalis, Stypandra glauca*) (Brown *et al.*, 2008). Studies on the chemical composition of floral fragrances emitted indicate that *T. macrophylla* is more closely mimicking *O. laxus* than does *T. crinita* (Edens-Meier *et al.*, 2014). Flowers of the *T. variegata* complex resemble those of *Calectasia* (Dasypogonaceae) (Fig. 7.18) although floral mimicry as a pollinator attractant has not been proven.

Diuris flowers are not usually sweetly scented nor do they offer a visiting insect any food reward. In *Diuris* the deception focuses on attracting a suite of pollinators from a guild of pea-flowered legumes (Fig. 8.6), thus increasing their evolutionary fitness by maximising insect visiting rates to reduce the risk of poor pollen transfer but increasing the potential for natural hybrids to occur between co-occurring species. Specifically the floral mimicry involves similarly coloured and shaped co-flowering peas (e.g. *Daviesia, Pultenaea, Isotropis, Gastrolobium, Gompholobium*) (Beardsell *et al.*, 1986; Indsto *et al.*, 2006; Brown *et al.*, 2008). Flowers of *Diuris purdiei* may mimic the shape, but not the colour, of co-occurring *Lobelia* (Campanulaceae).

Food deception in *Diuris* and *Thelymitra* has a moderate evolutionary cost because the requirements to select floral structures, colour, UV-guiding markers and positioning to precisely or generally mimic co-flowering species may be dependent on the evolutionary stability of floral traits and flowering periods of flowering species being mimicked. Mimicking flowers tend to be significantly larger than their models as a strategy to attract visiting insects at a short distance to the flowers after following a floral scent trail. This imperfect mimicry may not necessarily increase pollination success rates (Eden-Meier *et al.*, 2013).

8.1.3.3 Gnat Pollination

Other orchid genera attract fungus-eating gnats (e.g. *Corybas, Pterostylis* and *Rhizanthella*). *Rhizanthella gardneri* is a rare subterranean orchid restricted to two isolated populations in the SouthWest (Bougoure *et al.*, 2008; Chapter 6). Enlarged bracts (Fig. 8.7) that surround the underground inflorescence push through the top soil allowing the perfusion of a putrid odour that may mimic the scent emitted by fruiting bodies of mychorrizal fungi associated with the orchid's *Melaleuca* habitat.

Pterostylis flowers (Fig. 8.7) are complex traps that attract mosquitos and fungal gnats (Diptera). The labellum is hinged and moves in response to irritation by visiting insects that become trapped for 30–90 seconds inside the flower between the labellum and the androgynoecious column (Bernhardt, 1995). The labellum then spontaneously resests itself, allowing the visiting insect to escape. *P. sanguinea* is visited by the male of a *Mycomya* (Mycetophilidae) gnat species (Phillips *et al.*, 2014) attracted by pheromone-like semiochemicals emited from the flower's labellum.

Fig. 8.7: Fungal gnat pollinated orchids. *Rhizanthella gardneri (upper)* is obligately subterranean. Images of *C. discoidea* show that the tips of the enlarged fleshly bracts are the only part of the inflorescence that may protrude through the soil. In this image the soil surrounding the inflorescence has been removed to expose the 8-cm wide capitulum of deep red florets (6 mm in length), shown in close up on the right. (*lower*) *Pterostylis* (Greenhoods) often flower in autumn or winter when the gnats are most active (Brundett, 2007). *P. angusta* (*lower left*; flower length 30 mm) has the longest labellum of all SouthWest greenhoods that protrudes between two upright fused sepals. (*lower right*) *P. barbata* (bird orchid), flower length 50 mm, has a characteristic translucent flower and sparsely fringed labellum. *Rhizanthella* (except close up of florets) and *P. angusta* images provided by Mark Brundrett. *P. barbata* provided by Allen Lowrie.

8.1.3.4 Sexual Deception

Sexually-deceptive orchids are the most pollinator-specialised plants in the SouthWest flora (Phillips *et al.*, 2010), a syndrome that has evolved multiple times in phylogenetically different orchid genera (Mant *et al.*, 2002; Hopper, 2009). It is a strategy that tricks male insects to pseudocopulate with the labellum in a way that promotes pollen transfer. Initially arriving after following a scent trail of volatile chemicals, males utilise visual cues such as colour and modified labellum parts to be lured to the flower.

Sexually-deceptive *Caladenia* species attract male thynnine wasps (Phillips *et al.*, 2009), with the plant-wasp relationship species-specific or the wasp may pollinate closely-related species (e.g. *Zaspilothynnus nigripes* and the *C. huegelii* complex) (Phillips *et al.*, 2009). The male is attracted to an orchid by the volatile semiochemicals (pheromones) that the flower emits from either the labellum, or glandular petal and sepal tips, and visual cues such as the dark calli on the labellum of *Caladenia roei* (Fig. 8.6) that mimic the female. In the subgenera *Calonema* and *Drakonorchis*, floral parts are are highly engineered to attract their thynnine pollinators, including stiff- and loose-hinged labellae.

Drakaea, *Paracaleana* and *Spiculaea* (Fig. 8.8) rely on sexual deception by males of different thynnine wasp species that pseudocopulate with the column, resulting in the transfer of pollinia between wasp and orchid, and their flowers are highly modified to achieve this (Fig. 8.8). Labella are terminated by an appendage that visually mimic a wingless female (called the pseudofemale). *Drakaea*, *Spiculaea* and some *Caladenia* (subgenera *Drakonorchis)* have bioengineered passive labella with soft and hard hinges that utilise the attempted copulation of an attracted male to propel the insect towards the column as a mechanism to transfer pollen. In *Drakea*, the labellum contains a delicate hinge that allows the labellum to swing backwards towards the upright column as the male wasp attempts to fly away with the pseudofemale. Within *Paracaleana* the labellum relies on slight pressure to bend the labellum stalk (or claw) downwards 180° toward the downward-pointing column (Hopper & Brown, 2006; Bower, 2014). Pollen transfer is only effective if male wasps are precisely positioned onto the labellal pseudofemale, and only a small percentage of visiting males effect transfer (Peakall, 1990).

There is a high level of pollinator specificity in *Drakaea*, with each species pollinated by males of different *Zaspilothynnus* (Thynnidae) species (Hopper & Brown, 2007; Manz *et al.*, 2013). Pollen movement is largely restricted to plants within populations (Menz *et al.*, 2013). Pollination studies of *D. glyptodon* show that, although the majority (93%) of visiting males *Z. trilobata* wasps landed on flowers, less than 21% held onto the pseudofemale with only 6% attempting to copulate in a way that would lead to pollen transfer (Peakall, 1990). Labella are elongated with the tip modified to imitate the body of female thynnine wasps (Fig. 8.8). Analysis of the volitle compounds emitted by *D. glyptodon* has revealed a blend of pheromone-like pyrazines as an important component of the attractant (Bohman & Peakall, 2014; Bohman *et al.*, 2014).

Fig. 8.8: Wasp pollinated orchids. All species emit pheromone-mimicking chemicals and have labella apices modified to resemble female wasp bodies. From top left to lower right: *Drakaea glyptodon* (hammer orchid), *Paracaleana triens* (flying duck orchid), *Cryptostylis ovata* (slipper orchid) and *Spiculaea ciliata* (elbow orchid). Except for *Cryptostylis*, all genera are pollinated by male thynnine wasp species within the genera *Zaspilothynnus* (*Drakaea*; Hopper & Brown, 2007), *Eirone* or *Thynnoturneria* (*Paracaleana*; Hopper & Brown, 2006, Peakall *et al.*, 2010; Bower, 2014) or *Thynnoturneria (Spiculaea;* Alcock, 2010*)*. *Cryptosytlis* is pollinated by the male ichneumonid wasp, *Lissopimpla excelsa* (Gaskett, 2012). Flowers vary in length from 20 to 30 mm. Images provided by Allen Lowrie.

By targeting only one pollinating species, sexually-deceptive orchids maintain floral isolation (Schiestl & Schlüter, 2009) that promotes outcrossing and causes lower fruit set than other pollination strategies (Jersáková *et al.*, 2006; Phillips *et al.*, 2009), although this is not always the case (e.g. *Cryptostylis*; Gaskett *et al.*, 2008). It also has the potential to result in increased pollen movement (Peakall & Beattie, 1996). Among endemic SouthWest *Drakaea*, pollen movement distances of up to 132 m (*D. glyptodon* (Peakall, 1990), 556 m (*D. elastica*) and 267 m (*D. livida*) (Menz *et al.*, 2013) have been recorded, with pollen movement largely confined to within *Drakea* populations.

Hybridization in *Drakaea* is rare, occurring in mixed populations (Hopper & Brown, 2007), further supporting the secondary importance of colour mimicry and female wasp imitation for visits to occur. The evolutionary implications of extreme pollinator specialization will minimize the potential for hybridisation to occur, particularly between small, isolated populations, but makes the species more prone to local and regional extinction if the pollinator population begins to decline (Phillips *et al.*, 2010). The success of sexual deception not only requires optimal searching and movement between flowering individuals by males, but also the presence of co-flowering species that offer a food reward.

The only sexually-deceptive species that does not attract male wasps is the monotypic *Leporella fimbriata* that sexually deceives winged male ants of *Myrmecia urens* (Formicidae: Myrmeciinae) (Peakall *et al.*, 1990, Peakall, 1989; Fig. 8.9). Growing clonally from underground tubers the species is self-compatible as ants tend to pollinate within clones but minimises selfing by the elimination of selfed progeny (Peakall & James, 1989). About 60% of visiting ants result in successful pollination (Peakall, 1989).

Fig. 8.9: A male winged *Myrmecia urens* ant positioning itself sideways along the wide labellum of the sexually deceptive *Leporella fimbriata*. Labellum is 10 mm wide. Image provided of Mark Brundrett.

In evolutionary terms, sexual deception may be viewed as a 'high cost' strategy as each orchid species obligately deceives a specific pollinating insect, although there is potential for more than one wasp species to pollinate widespread species, as may be the case for sexually-deceptive caladenias (Phillips *et al.*, 2009). It has been argued that the modified floral traits of sexually deceptive orchids have a minimal influence on their pollinator's evolution (Schiestl, 2005; Jersáková *et al.*, 2006). This is certainly the case for *Cryptostylis* (Fig. 8.8) where all five Australian species are only pollinated by the wasp *Lissopimpla excelsa* (Ichneumonidae), emitting the same chemical attractant (Schiestl *et al.*, 2004) and exhibiting close-range colour mimicry (Gaskett & Herberstein, 2010) but it does not account for the unparalled diversity and specificity of Thynnine wasp species that pollinate *Drakaea* and sexually-deceptive species of *Caladenia* (Hopper & Brown, 2007; Phillps *et al.*, 2009). Orchid speciation may be a result of subtle changes in the chemical structure of the volatile semiochemicals emitted as a response to speciation rates within their target wasp genera, as has been predicted for *Chiloglottis* species inhabiting the temperate regions of southeastern Australia (Mant *et al.*, 2002; Peakall *et al.*, 2010). Since *Zaspilothynnus nigripes* is the pollinator of *Drakea livida* and several species within the *Caladenia huegelii* complex (Hopper & Brown, 2007; Phillips *et al.*, 2009; Manz *et al.*, 2013), these species probably emit an analogous blend of semiochemicals as their wasp attractant.

8.2 Honeyeater-pollinated Species

Honeyeaters (Meliphagidae) (Fig. 8.10; 8.12) are the main driver of bird-pollination floral traits in the SouthWest flora and the Australian flora in general. Early divergence of honeyeaters occurred 23.5 Ma and radiated strongly between 15 and 5 Ma (Joseph *et al.*, 2014), coinciding with the evolution of the bird-pollinated genus *Leptosema* and radiation of bird-pollinated *Gastrolobium* and *Bossiaea* species (Toon *et al.*, 2014) and hakeas (T. He, unpublished). This represented an evolutionary shift from bee-pollination to bird-pollination syndromes in pea-flowered legumes (Fabaceae, tribes Mirbeliae and Bossiaeeae) and hakeas. Similarly, the honeyeater-pollinated genus *Anigozanthos* (Haemodoraceae) originated 14 Ma from insect-pollinated relatives (Hopper *et al.*, 2009) (Fig. 1.1).

Species targeting honeyeater pollinators typically display relatively large, prominent, red-coloured flowers and inflorescences (Fig. 8.10, 8.11) that utilise the bird's ability to discriminate more towards the red end of the visible colour spectrum (Shrestha *et al.*, 2013), and thus are easily visible among the green foliage and post-fire blackness of the SouthWest vegetation. Prominent flower and inflorescence positioning, including pendulous flowers, have repeatedly evolved in phylogentically-independent groups to accommodate honeyeaters that need to perch while nectar feeding. An exception to this are the flower heads of *Banksia* (excluding subgenus *Dryandra*) that attract an array of nectar-seeking insects as well as honeyeaters (Fig. 8.12). As there are few red-flowering *Banksia* species (e.g. *B. menziesii*; Fig. 7.6; *B. coccinea*, Fig. 8.11), honeyeaters visits are a result of learnt foraging behaviours that maximise nectar rewards, with colour-change cues differentiating between unopened and nectar producing florets (Fig. 7.6; 8.12). Honeyeaters feed from the yellow-flowered *B. attenuata* but are not considered its primary pollinators (Wiens *et al.*, 1979; Whelan & Burbidge, 1980).

Fig. 8.10: Examples of honeyeaters that are common in, but not necessariy restricted to, the SouthWest. Brown honeyeater (*Lichmera indistincta*), singing honeyeater (*Lichenostomus virescens*) and white-cheeked honeyeater *(Phylidonyris nigra)*. All genera arose 6–8 Ma (Joseph *et al.*, 2014). Birds pictured are 14–20 cm in length, the brown honeyeater being the smallest honeyeater in the SouthWest. (*lower left*) The floral tube of *Anigozanthos rufus* (Haemodoraceae) is elongated and partially fused as a means of ensuring only honeyeaters are cabable of accessing the nectar reward, whereas in *Eucalyptus forrestiana* (Myrtaceae) (*middle*) the upright yellow stamens create a tunnel-like barrier to nectar seeking insects, but allows access to visiting honeyeaters. (*right*) *Leptosema daviesioides* (Fabaceae), known as the upside-down pea bush, produces large red flowers from the base of the plant (geoflorous). All flowers 4 cm long. Honeyeater images provided by Geoffrey Groom; floral images by *Esperance Wildflowers*.

Templetonia retusa
(Fabaceae)

Kennedia prostrata
(Fabaceae)

Leptosema aphyllum
(Fabaceae)

Hakea orthorrhyncha
(Proteaceae)

Hakea laurina
(Proteaceae)

Hakea francisiana
(Proteaceae)

Adenanthos barbiger
(Proteaceae)

Banksia coccinea
(Proteaceae)

Lambertia multiflora
(Proteaceae)

Calothamnus graniticulus
(Myrtaceae)

Eucalyptus rhodantha
(Myrtaceae)

Anigozanthos manglesii
(Haemodoraceae)

Fig. 8.11: Selected examples of bird-pollinated (honeyeaters) flowers in the SouthWest flora showing their great structural diversity. The larger flowers/inflorescences (*B. coccinea, E. rhodantha, H. francisiana, H. laurina*) may also be pollinated by small marsupials such as the honey possum.

SouthWest genera that are considered obligate meliphagiophilous (honeyeater-pollinated) ensure minimal pollen loss to other animal vectors by positioning anthers and stigma in the same location (3–5 cm from the nectar source), even if out-of-phase *via* protandry, to maximise pollen transfer when the bird visits another flower (e.g. *Anigozanthos, Leptosema, Calothamnus*; Fig. 8.10, 8.11). Other species exclude insect visitors from accessing nectar by imposing impenetrable floral barriers, e.g. *Eucalyptus stoatei* (Hopper & Moran, 1981) and the related *E. forrestiana* (Fig. 8.10), or shield the flowers with conspicuous but dull bracts (e.g. *Pimelea physodess*, Fig. 8.12).

Brush-shaped honeyeater-pollinated flowers are common within the families Myrtaceae, Proteaceae and Fabaceae (Faboideae) (Keighery, 1980) where the pollen is 'brushed' all over the bird's beak and head. Gullet-shaped flowers occur in *Calothamnus* (Myrtaceae), *Anigozanthos* (Haemodoraceae) and some species of *Templetonia* (Fig. 8.11). In all instances visiting birds must reach the nectar at the base of the flower, and in doing so brush pollen onto their heads and beaks.

Distinct bird-pollination syndromes also occur within the Ericaceae (Johnson, 2013). In *Daviesia* (Fabaceae) only two closely-related species are honeyeater pollinated (*D. epiphyllum* (Fig. 9.5) and *D. speciosa*) (Crisp & Cook, 2003, Toon *et al.*, 2014). A full list of bird-pollinated species is provided in Keighery (1980).

Fig. 8.12: Red wattlebird (*Anthochaera carunculata*) (*left*) sourcing nectar from a pendulous *Pimelea physodes* (Thymelaeaceae) inflorescence, the only bird-pollinated *Pimelea* (Keighery, 1975). Flower head 5 cm long. Western wattlebird (*Anthochaera lunulata*) (*right*) using the inflorescence of *Banksia speciosa* as a landing platform, feeding from recently opened nectar-producing florets. Western Wattlebird image provided by Lochman Transparencies.

8.2.1 Case Study: *Gastrolobium*

Gastrolobium celsianum (synonym *Brachysema lanceolatum*) (Fig. 8.13) is in the most advanced clade of gastrolobiums and is bird-pollinated (rare among peas) (Crisp, 1996; Chandler *et al.*, 2002; Toon *et al.*, 2014). The clue is the large red flowers, with small standard and wings and abundant nectar. When probed from above while seeking nectar the bird depresses the keel and initially exposes the anthers (Fig. 8.13, *upper right*) that brush their pollen against the nape and beak and later the receptive stigma receives pollen from the same part of the bird (protandry). Coverage by the keel prevents rainwater washing away the presented pollen. The less advanced species, *G. bilobum*, has dense racemes of small, yellow, insect-pollinated flowers with the standard exceeding the length of the keel. The terminal inflorescence is vulnerable to florivores, such as emus, that have evolved however to withstand the exceptionally high levels of the toxin fluoroacetate, up to 2 mg g^{1}, that they contain (Twigg *et al.*, 1996; Twigg & Socha, 1996). It is unknown if *G. celsianum* also contains fluoroacetate but the red coloration suggests that they are more likely to contain cyanogenic glycosides to which immunity is unlikely and is a feature of red-coloured, bird-pollinated hakeas and grevilleas for example (Hanley *et al.*, 2009).

Fig. 8.13: *Gastrolobium celsianum* (*upper*) is bird-pollinated, displaying large red flowers, 35 mm long. (*lower*) *G. bilobum* is insect-pollinated, flowers 8 mm long.

8.2.2 Pollen Dispersal

An advantage of utilising honeyeater as pollinators is their ability to carry significant pollen loads over long distances within relatively short time frames (Byrne *et al.*, 2007), although honeyeaters tend to move more frequently between inflorescences on the same plant (Wooller *et al.*, 1983; Collins & Spice, 1986; Day *et al.*, 1997; Byrne *et al.*, 2007). It has been estimated that for hybridisation to have occurred between *Banksia hookeriana* and

B. prionotes the birds must have carried successful pollen for a distance of at least 250 m (Lamont *et al.*, 2003). This is particularly advantageous for trees and shrubs either restricted to specific habitats (e.g. granite outcrops, heathlands) and species with disjunct and fragmented populations. Some prolific post-fire flowering species (e.g. *Anigozanthos* species, *Verticordia grandis*) take advantage of the barren, burnt landscape to attract honeyeaters (Lamont & Downes, 2011). This maximises the incidences of pollen load transfer between flowers within and between populations as honeyeaters need to visit more flowers per species because other species are not in flower. Floral visiting rates are high as honeyeaters continually need to frequently visit nectar-producing flowers to meet their daily energy requirements (Collins & Morellini, 1979; Collins, 1981; Collins & Briffa, 1983; 1985). Honeyeaters utilise multiple plant species, the composition depending on seasonal availability of nectar (Hopper, 1981; 1980), sourcing nectar from adjacent habitats as necessary (Collins, 1985).

Studies on two iconic granite outcrop endemics, *Kunzea pulchella* and *Eucalyptus caesia* (Myrtaceae), have shown that they have a history of population isolation and persistence from at least the Pleistocene. Both possess prominent nectar-laden red flowers that are pendulous in *E. caesia* (see Fig. 7.3 for photograph of *E. caesia* flower). Honeyeater pollination between neighbouring outcrops is a driver of population divergence caused by genetic drift (Moran & Hopper, 1983; Byrne & Hopper, 2008; Tapper *et al.*, 2014). The reliance on honeyeaters is most important for the geographically restricted *E. caesia* because of its inability to inbreed (Byrne & Hopper, 2008) and its protrandrous nature. Bird pollination maintains genetic diversity of populations by promoting outcrossing, particularly those that are disjunct and fragmented. This is the case for the rare *E. rhodantha,* despite its low level of outcrossing, where high levels of gene flow between populations maintain relatively high levels of population heterogeneity (Sampson *et al.*, 1989) and also in the outcrop endemic *Kunzea pulchella* (Tapper *et al.*, 2014).

8.3 Mini-possum Pollination

The main marsupial pollinators (therophily) of the SouthWest flora are the minute honey possum (*Tarsipes rostratus*: Tarsipedidae) and western pygmy possum (*Cercartetus concinnus*: Burramyidae) (Fig. 8.14), although rodents (e.g. bush rat, *Rattus fuscipes:* Muridae) may also be active. Mini-possums are the most recently evolved pollinators of the SouthWest flora. Weighing up to 16 g, both species are twilight-active and nocturnal opportunistic foragers for suitable nectar producing and pollen bearing flowers. *T. rostratus* is endemic to the SouthWest and a nectar-feeding specialist, the only non-flying mammal to do so, with a 'brush' tongue for lapping up nectar and pollen (Richardson *et al.*, 1986). *C. concinnus* is omnivorous, consuming fruit, seeds, insects and spiders. These mini-possums forage almost exclusively on species with large inflorescences and flowers within the Proteaceae, Myrtaceae and Ericaceae (Hopper,

1980; Bradshaw *et al.*, 2007; Pestell & Petit, 2007; Dundas *et al.*, 2013), widespread in the flora of their heath and woodland habitats. Many of the species visited and pollinated by these mini-possums are also visited by honeyeaters (e.g. *B. coccinea, B. baxteri, B. occidentalis)* (Hopper, 1980). Mini-possums may be deterred from visiting the large pink-red flowers of bird-pollinated *Hakea* (Proteaceae) species because of the lack of floral scent and presence of floral toxins (cyanogenic compounds) (Hanley *et al.*, 2009). Mammal-pollinated inflorescences are usually not brightly coloured, hidden within the foliage (cryptic), have a strong, often musky, scent and produce copious viscous nectar that may run to the ground as a scent trail.

Fig. 8.14: (*upper*) Honey possum (*Tarsipes rostratus*), an obligate nectarivore, feeding from *Banksia* inflorescences (*left*) about to probe *B. benthamiana* that does not present its pollen, (*right*) probing 20-cm-long flower head of the rare *B. tricuspis* that presents its pollen on hooked styles so that pollen is combed through the possum's fur. It is a highly effective pollinator of *B. tricuspis*. (*bottom*) Western pygmy possums (*Cercartetus concinnus*) feeding on pollen and nectar from flowers of *Eucalyptus preissiana*, 5 cm wide. Because of the many stamens, pollen is deposited on their snouts, heads and ears. One individual is also using its forelimb to collect pollen. Image of *B. tricuspis* by Stephen van Leeuwen. Other images provided by Lochman Transparencies.

Primarily restricted to a home range up to only 0.8 hectares, with males foraging wider than females, it has been sugested that *T. rostratus* has an intimate knowledge of the local food sources available throughout the year (Garavanta *et al.*, 2000; Bradshaw & Bradshaw, 2002). Nectar is scarce during the dry summer months compared with the winter months (Wooller *et al.*, 1984; Bradshaw & Bradshaw, 1999), with some individuals recorded to move as far as 400 m in search of a suitable food source (Bradshaw *et al.*, 2007).

T. rostratus is the predominant pollinator of many *Banksia* species with hooked styles. Their flowering heads contain hundreds of florets each with recurved styles terminated by a pollen presenter. This transfers pollen to and from the possum's fur as they clamber in search of nectar (Fig. 8.14). Mini-possum-pollinated banksias are typically shrubby and short in stature or prostrate, although both mini-possums have been observed feeding from species that are small trees. Arborescent banksias pollinated by mini-possums (e.g. *B. nutans*, *B. grandis*, *B. tricuspis*) all have pale yellow or brown flower heads often held within the crown rather than displayed prominently as in bird-pollinated species (Table 8.1). Interestingly, disjunct populations of the yellow-flowered *Banksia verticillata*, restricted to granite outcrops on the south coast of the SouthWest, are primarily pollinated by honeyeaters that promote outcrossing. Some mini-possum-pollinated species previously ascribed to the now defunct genus *Dryandra* are ground flowering (e.g., *B. brunnea*, *B. proteoides*; Fig. 7.13) and cryptic (e.g. *B. nivea*, *B. dallanneyi*, *B. plumosa*). Prostrate banksias (*B. goodii* complex) are also ground flowering (geoflorous) with rusty-brown, possum-pollinated, flower heads (Fig. 2.6). It has been speculated that the geoflorous habit in *Banksia* is derived from bird-pollinated ancestors as a means of avoiding the impact of low intensity fires (Rourke & Wiens, 1977) but these species in fact are highly vulnerable to ground fires and they also retain their dead foliage and florets that promote their flammability (He *et al.*, 2011). *Banksia nutans* is a 1-m-tall shrub with pendulous, cryptic, purple-brown, musky, nocturnal-opening flower heads (Table 8.1) whose nectar may drip to the ground and is exclusively pollinated by *T. rostratus* (Wooller & Wooller, 2003). This species has the ability to self-pollinate, a strategy to maintain seed production in the absence of honey possum visits (Wooller & Wooller, 2003).

Mini-possums are unlikely to have had a major evolutionary influence on plant trait adaptation in the SouthWest due to their small size and limitations on their daily energy requirements for travelling long distances. There is also significant competition for limited seasonal nectar resources with the daylight-active, and more mobile, honeyeaters. Nevertheless, some plant species, such as *B. nutans*, have adapted strongly to their ongoing presence over the last few million years.

9 Leaf Properties

The following quote typifies the variability among SouthWest leaves:

> "*That plants of the south-west should exhibit a variety of leaf types is not surprising. The features of interest are that:*
>
> 1) *Such widely divergent leaf types occur within single genera, and evidently most of them represent autochthonous* [occurring at its place of origin] *development;*
> 2) *Some of these leaf types are unique, not found elsewhere;*
> 3) *With all genera compared, there is such a high degree of convergent evolution; and*
> 4) *There is apparently easy reversibility in morphology.*
>
> *Leaf morphology, a rapidly evolving feature, can exhibit sensitive adaptation for a particular habitat, although it may in some cases lag behind physiological adaptation, and various leaf types are equally successful in a given locality*".
>
> Sherwin Carlquist (1974) *Island Biology*. Columbia University Press, New York

The SouthWest flora has an exceptional array of leaf shapes and sizes, including phyllodes and cladodes. Leaf (or leaf-like) diversity is greatest among the most speciose shrub genera including *Melaleuca*, *Acacia*, *Grevillea*, *Hakea*, and *Daviesia*. Note that the *Melaleuca* complex now incorporates the SouthWest endemic genera *Beaufortia*, *Calothamnus*, *Conothamnus*, *Eremaea*, *Phymatocarpus* and *Regelia* (Edwards *et al.*, 2010; Craven *et al.*, 2014). The level of diversity is especially high in the hyperdiverse sandplain floras where co-occurrence is driven by niche differentiation, habitat filtering (species with similar ecological requirements co-occur more often than expected by chance), biogeographic history and phylogenetic clustering (Merwin *et al.*, 2012). In addition, many adaptive solutions to the same soil nutrient levels, light intensities, air temperatures, winds and moisture regimes are feasible, creating a rich tapestry of form and function.

9.1 Leaf Size

Surviving the hot, dry summers of the SouthWest requires structural and physiological strategies that maximize leaf water-use efficiency (see Chapter 3) without compromising leaf heat resistance. These include a reduction in leaf size, an increase in mechanical strength (thicker cells, more fibrous tissue) and changes in leaf absorbance properties (e.g. presence of light-reflective hairs). The ability of a species to tolerate extreme summer temperature events will be determined by the interaction between leaf heat loads (*i.e.* rise in leaf temperature, a function of leaf size and transpiration rate), heat-storing capacity (a function of leaf thickness) and the degree of exposure to the prevailing environmental conditions (Groom *et al.*, 2004).

Most SouthWest scrub-heath species have leaves that can be categorised as leptophyllous (< 25 mm^2 in one-sided area), the smallest size class recognized (Table 9.1). Small areas both minimize leaf water loss and reduce the likelihood of excessive, and potentially lethal, leaf temperatures occurring during the hot summers. Leaf size and level of sclerophylly can vary considerably within a genus and even within a species from juvenile to adult leaves—heteroblasty (e.g. *Hakea*; Groom *et al.*, 1997). *H. trifurcata* is unusual in its seasonal production of broad and needle leaves (seasonally recurrent heteroblasty) (Groom *et al.*, 1994). Eucalypt woodlands and forests of the SouthWest have a higher percentage of broad-leaved species present in their understorey than the sandplain shrublands because they experience less exposure (more shade).

Table 9.1: Leaf traits (%) of 429 vascular plant taxa (including 2 gymnosperms, 75 monocotyledons, 352 dicotyledons) at four sites containing kwongan vegetation (250 km N of Perth). Kwongan is an indigenous term for the low, heath-like sclerophyllous sandplain shrublands. Phyllodes and phylloclades are treated as leaves here though they are more like green stems in structure. Data collated from Pate *et al.*, (1984).

Leaf shape				
Scale leaves	Needle-like	Strap-like	Dissected flat	Broad flat
11.4	26.3	36.6	8.3	17.5
Leaf size category (cm^2)				
Leptophyll	Nanophyll	Nano-microphyll	Microphyll	Mesophyll (and larger)
<0.25	0.25–2	2–12	12–56	56–180+
54.3	21.4	18.6	5.0	0.4
Leaf consistency				
Sclerophyll (hard)	Malacophyll (soft)	Resinous	Succulent	No leaves
68.0	27.5	0.9	3.5	0.2
Leaf colour				
Green	Green and white	White, hoary	Glaucous (grey, powdery)	Brown
69.7	8.1	1.9	10.7	9.6
Leaf life span (years)				
< 1	1	2–3	4–5	> 5
20.7	40.8	35.6	1.9	1.0

Fig. 9.1: Variability in leaf size, shape and texture of the shrub and tree species in the SouthWest flora. Species are (from top to bottom, left to right: Margins are sharply mucronated (pointed) on the leaves of *Hakea amplexicaulis* (Proteaceae), *Banksia grandis* (Proteaceae), *Hakea horrida, Grevillea bipinnatifida* (Proteaceae), *Melaleuca huegelii* (Myrtaceae), *Calothamnus sanguineus* (Mytaceae) and *Hakea psilorrhyncha* (Proteaceae) whereas the soft compound leaves of *Acacia pulchella* (Fabaceae) are subtended by sharp thorns (photo shows the pinnae of *A. pulchella* bipinnate leaves folded together in response to the summer drought). Leaves vary in length from 15 cm in *H. amplexicaulis* to < 5 mm in *M. huegelii* and in thickness from 0.5 mm in *A. pulchella* to 3 mm in *H. psilorrhyncha*.

9.2 Leaf Design

In Mediterranean-type climates the degree of sunlight exposure and water accessibility during the hot, dry summer months are strong abiotic selective forces on the leaf design of extant species. Trade-offs between photosynthetic capability, water-use efficiency and leaf heat loading control leaf size and shape constrained by phylogenetic and lineage limits (Ackerly, 2004). Many SouthWest species inhabit landscapes with minimal or no shade and are therefore exposed to prolonged hot and dry summer conditions, and this has implications on leaf design, leaf arrangement on the stem (phyllotaxy) and leaf orientation.

9.2.1 Reducing Summer Leaf Heat Loads

Leaf shape and size are linked to photosynthetic rates and leaf thermotolerances (Knight & Ackerly, 2002; 2003; Nicotra *et al.*, 2008). Taxa with highly dissected, pinnate, lobed and serrated leaves (Fig. 9.1) typically have a greater photosynthetic tolerance of high temperatures, and dissipate heat quicker than if the leaf was entire (Nicotra *et al.*, 2008), and thus are favoured in landscapes exposed to periods of prolonged high air temperatures. Where leaves (or phyllodes) are not dissected or lobed they may be orientated vertically as an alternative means of reducing heat loads (Smith *et al.*, 1998) and protecting the photosynthetic process against excessive leaf temperatures when stomates close as a response to summer water stress.

Approximately a quarter of the SouthWest's woody shrub species possess needle-like leaves that are either single (e.g. *Hakea psilorrhyncha*) or highly dissected (e.g. *H. horrida*) (Fig. 8.1), and are typically highly spinescent at their apices. Leaf spines may not only serve as an anti-herbivore or granivore deterrent, but also assist in the dissipation of summer leaf heat loads. Needle leaves have a reduced surface area and an increased leaf thickness compared with broad leaves. Clusters of thick, minute (leptophyll-nanophyll) leaves (e.g. *Melaleuca huegelii*, *Calothamnus sanguineus*, Fig. 9.1) are another mechanism to minimize leaf heat loads. Thick leaves retain more thermal energy than co-occurring thinner leaves (Groom *et al.*, 2004) and have a lower rate of change in leaf temperature (Jones, 1992). Leaf thickness has a dampening effect on heat absorption and irradiance damage avoidance (Leigh *et al.*, 2012). The impact of increased thickness is greatest when other means of reducing leaf temperature (transpiration, reflectance or reduced size) are lacking (Leigh *et al.*, 2012).

Needle-like leaves first appeared in SouthWest *Banksia* and *Hakea* species 14–15 Ma (He *et al.*, 2011; T. He, unpublished data) during a period of aridification that marked the beginning of today's Nullarbor Plain. Extant needle-leaved species of the widespread genus *Hakea* are more likely to be found in landscapes with less reliable rainfall (the Wheatbelt and Mallee woodlands and heaths of the Esperance Plains), whereas broad-leaved species tend to occur where there is more reliable seasonal rainfall (e.g. Geraldton sandplains) or less extreme annual temperatures (e.g. the forested regions) (Groom & Lamont, 1996).

9.2.2 Heterophylly

The Southwest flora is noted for the strong presence of heterophylly (different leaf types within a plant), mainly as a heteroblastic process with the juvenile leaves tending to be more mesomorphic and the adult leaves highly scleromorphic (Groom *et al.*, 1997; Zotz *et al.*, 2011). Heterophyllous traits may have evolved as part of a mimicry syndrome (e.g. *Hakea trifurcata;* Fig. 3.5) or they may serve to protect the plant directly from florivores and granivores when they flower and fruit (e.g. *H. auriculata*; Fig. 9.10; 9.11). In both cases, the highly-modified, protective leaves only appear once the plant starts to flower (Groom *et al.*, 1994). In the Fabaceae, heteroblasty involves the initial production of simple or compound leaves transitioning to phyllodes or cladodes (Walker & Pate, 1986; Forster & Bonser, 2009), e.g. *Daviesia horrida*. In phyllodinous *Acacia* species (Fig. 9.2), the ontogenetic transition from (bi)pinnate compound leaves to phyllodes in the early seedling stage is viewed as evolutionarily redundant, reflecting the leaf form ancestry of the genus (Gardner *et al.*, 2008). This includes a transitionary phase where successive leaves are produced with an expanding flattened petiole and a gradually reducing lamina of terminal leaflets until only the phyllode is formed (Kaplan, 1980; Fig. 9.2).

For species within section *Pulchellae* (e.g. *Acacia pulchella*; Fig. 9.2) the first two true leaves are pinnate and formed simultaneously, with opposite phyllotaxy (Guinet *et al.*, 1980; Miller *et al.*, 2003). Ontogeny whereby a pinnate leaf develops first, followed by bipinnate leaves, occurs in a range of taxonomic subgroups, including the majority of section *Acacia* (Miller *et al.*, 2003; Murphy *et al.*, 2010; Miller & Miller, 2011; Miller *et al.*, 2013). Heteroblasty in *Acacia* is considered non-adaptive (functionally neutral) because of the short timeframe between the formation of true leaves and phyllodes, all of which are present together in the first growing season (Zotz *et al.*, 2011; Fig. 9.2).

Fig. 9.2: (*left*) A heteroblastic *Acacia* (Fabaceae) seedling displaying the transition from pinnate leaves (1), to an intermediate stage (2) where compound leaves are subtended by a vertically-flattened petiole to the formation of phyllodes only (3). Seedling is 25 cm tall. (*right*) Selection of 'leaves' within the genus: (*top*) bipinnate leaves of *A. pulchella* each basal to a 12-mm-long thorn. This is a member of section *Pulchellae* that consists of species mostly in Western and Northern Australia that radiated 20–22 Ma (Miller *et al.*, 2013). (*middle*) Vertically-orientated phyllodes of *A. camptoclada* with a prominent central vascular vein (uninerved) that ends in a blunt spine. Phyllode 15 mm long. (*lower*) Phyllodes of *A. cochlearis* with prominent 3-4 raised veins (trinerved). Phyllode 35 mm long. Photos of *A. camptoclada* and *A. cochlearis* provided by *Esperance Wildflowers*.

9.3 Sclerophylly

Sclerophylly refers to how tough, stiff or leathery a leaf is. Various indices of sclerophylly have been proposed. However, there is one simple measure that incorporates both morphological and structural aspects in our understanding of sclerophylly. This is leaf mass per area (LMA) that relates to how much structure (e.g. sclereids, cell wall thickening, palisade layers) is invested per unit area of projected (one-sided) leaf area (Groom & Lamont, 1999). The more a leaf invests in strength and support, the more sclerophyllous it will be, resulting in an increase in LMA (Read *et al.*, 2005). LMA is the product of leaf thickness and leaf density, representing different structural properties of the leaf (Witkowski & Lamont, 1991; Niinemets, 2001; Poorter *et al.*, 2009). The level of leaf toughness can also be measured directly and involves determining the amount of force required to fracture a leaf.

Southwestern Australia has some of the most nutrient-impoverished soils in the world (Lamont, 1995; Groom & Lamont, 2010). This can be attributed to its history of geological stability over the last 100 million years with continual leaching of the soil profile under generally high rainfall until 15 million years ago without landscape rejuvenation through tectonic activity. Thus, it is not surprising that nutrient content of the leaves of evergreen species here is also among the lowest in the world (Wright *et al.*, 2004). Thus, leaves of hakeas and banksias have a P content of 0.025% and N content of 0.55% on average compared with the average for floras in the rest of the world of 0.10% and 2.5% respectively (Lamont *et al.*, 2002; Witkowski & Lamont, 1996; Wright *et al.*, 2004). The low nutrient contents of the leaves is partly attributable to their high internal density because they are so sclerophyllous: mean LMA of 440 $\mu g\ mm^{-2}$ versus mean for floras elsewhere in the world of 130 $\mu g\ mm^{-2}$. That is, the nutrients are diluted by the high levels of structural material per unit volume of leaf.

Why are the trees and shrubs so sclerophyllous? In a low nutrient environment, an effective strategy is to have long-lived leaves so that absorbed inorganic nutrients can be re-used many times even when the metabolic rates are low. But the longer leaves survive the greater the chances that they will succumb to herbivores, drought, temperature extremes and wind buffeting. This is where the two components of leaf-mass-area as an index of sclerophylly become relevant. Incorporation of extra cellulose and lignin into the tissues increases rigidity of the leaf, improving support, preventing collapse during drought, reducing palatability and deterring herbivores directly *via* hardened edges and spiny teeth or apices (Figs. 3.4, 3.5, 9.3; Hanley *et al.*, 2007). Just as cell walls are thickened so cells are smaller and the greater the ability to conduct sap through the leaf (Jordan *et al.*, 2013). Thicker cuticles prevent water loss *via* the epidermis (cutin is hydrophobic) and they may even envelop the stomates.

Thickening the leaf increases light capture *via* the multilayed palisade, reduces transpiration rates by increasing resistance to movement of water from the xylem to the boundary layer, and increases its ability to store water in the central parenchyma and cell walls (Lamont & Lamont, 2000). Deposits of calcium oxalate, silica, tannins, starch, the amino acid proline and anthocyanins increase with leaf age and supplement the ability of leaves to deter herbivores, resist pathogens and tolerate drought through osmotic adjustment (Richards & Lamont, 1996). All may be involved in the greater survival of sclerophyllous species to abnormally high temperatures (Groom *et al.*,

2004). Increase in sclerophylly is accompanied by a reduction in leaf size associated with smaller cell size, vertical leaf orientation and conversion of dorsiventral leaves to isobilateral (stomates on both sides), frequently to a needle-shape (isolateral). These reduce the heat load on the leaf even when transpiration is negligible but allow photosynthesis to occur at the cooler time of day in warm weather when light is low on the horizon at dawn and dusk.

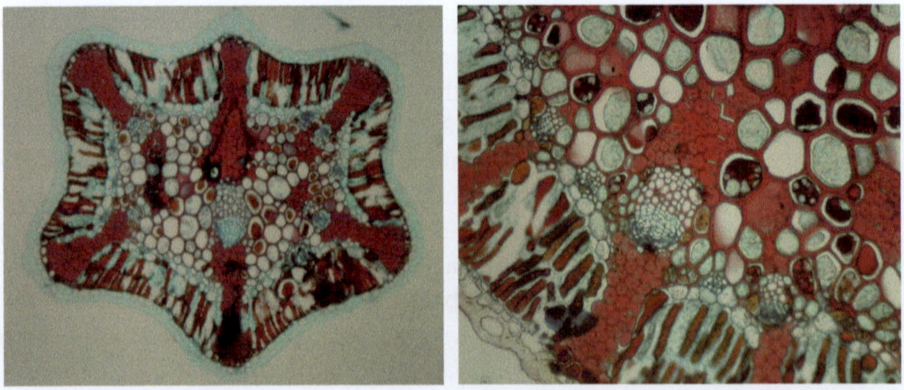

Fig. 9.3: The genus *Hakea*, with over 100 taxa in the SouthWest, contains the world's most sclerophyllous species (Wright *et al.*, 2004). Pictured are tranverse sections of the needle leaves of *(left) H. gilbertii* and *(right) H. subsulcata* that are widespread in the wheatbelt and mallee regions of the SouthWest (Chapter 1). The sections have been treated with fast green that stains cellulose, cytoplasm and cutin green and phloem blue, and safranin that stains liginin and tannin red and cutin pink. Conspicuous are the 'girders' of lignified fibres that strengthen the leaf and prevent collapse of the delicate chlorophyll-bearing double palisade during severe drought. The girders are topped by a tannin-bearing hypodermis that restricts light uptake and hence the heat load on the most exposed ridged parts of the leaf (Jordan *et al.*, 2005). There are also inner fibre groups wrapped around the vascular bundles that restrict water loss from the xylem. The inner parenchyma has walls thickened with lignin that increase the water-storing capacity of the leaf. Both the parenchyma and palisade have their vacuoles loaded with tannin that increases their osmotic concentration and assists in water retention through osmotic adjustment (Richards & Lamont, 1996). The cuticle is 2–3 times thicker than the epidermal cells and it rises over the sunken stomates (completely covering the opening) to form epistomatal cavities, sometimes with mucilaginous plugs in the substomatal cavities, all serving to restrict water loss further. These features help to explain why leaf-mass-area, as an index of sclerophylly, is higher among hakeas the lower the rainfall in the SouthWest independent of the N and P contents of the leaves (Lamont *et al.*, 2002).

In summary, the increasingly nutrient-impoverished soils in the SouthWest promoted the evolution of long-lived leaves. This was achieved through morphological and anatomical changes centred on sclerophylly that involved improved leaf support and herbivore and pathogen deterrence at first. Then, as plants became more exposed to light, heat and drought, and more efficient herbivores, as they moved out of dense forests over geological time, further structural and physiological changes evolved indicative of greater tolerance to high light and water deprivation. All these contributed to the extreme levels of scleromorphy present in the current flora, as LMA follows both nutrient gradients (Read *et al.*, 2005) and rainfall gradients (Lamont *et al.*, 2002) in the SouthWest. Spinescence also increases with decreasing rainfall effectiveness (B. Lamont, M. Hanley, P. Groom & T. He, unpublished data).

9.3.1 Case Study: Two Banksias with Long-Lived Leaves

Banksia petiolaris (Proteaceae) is a ground creeper with large vertical leaves while *Banksia baueri* is a shrub with small horizontal leaves (Fig. 9.4). They co-occur in scrub-heath between Albany and Esperance with an annual rainfall of about 500 mm. The sand-over-laterite substrate is nutrient-impoverished, with available P of only 2 ppm (Witkowski *et al.*, 1992). The leaves of *B. petiolaris* survive for up to 13 years and those of *B. baueri* for up to 5 years, both exceptionally long compared with non-sclerophyllous species. Leaf–mass–area (LMA) is similar in both species, except *B. petiolaris* leaves are 45% thicker than *B. baueri*, and *B. baueri* leaves are 40% denser than *B. petiolaris*. LMA in *B. petiolaris* increases from 250 μg mm^{-2} in the first year to 380 μg mm^{-2} in the eleventh year entirely due to an increase in leaf density over that time. Leaves of both species remain physiologically active throughout their long lives, maintaining strong control over their xylem water potentials, rates of transpiration and photosynthesis on a leaf area and nutrient basis, and water-use efficiency, though at low levels compared with non-sclerophylls, and lower still in the more sclerophyllous *B. petiolaris*, with all variables declining with leaf age.

Fig. 9.4: (*left*) *Banksia petiolaris* showing its stoloniferous habit and erect leaves, 20–25 cm long, arising at ground level and known to survive for up to 13 years. (*right*) *B. baueri* is a 1.5 m tall shrub with 5–10 cm long leaves arranged at a slight angle to the horizontal that survive for up to 5 years apparently cut short by self-shading. Both species retain their dead florets and leaves that enhance plant flammability and ensure follicle opening and seed release, important as both species are killed by fire.

9.4 Leaf-like Structures

Some genera do not produce true leaves, or they are highly reduced or modified. When the leaves are highly reduced, either the stem or the petiole takes on the photosynthetic function (Fig. 9.5). It is probably no accident that all the genera noted here contain above-average levels of nutrients in their foliage because of root adaptations to enhance nitrogen uptake (see Chapter 6) or are parasitic (Chapter 5), increasing their attractiveness to herbivores. All these modifications can be considered drought adaptations where survival is at the expense of growth rates.

1. *Acacia* (Fabaceae: Mimosoideae) – Adult leaves in this genus are typically either bipinnately compound (rare in SouthWest), phyllodinous, or reduced to scales. Phyllodes are expanded petioles that usually grow vertically rather than twist vertically as in true leaves. In some species (e.g. *A. stenoptera*) the phyllode continues down the stem to give a winged structure collectively called a phylloclade. In other species these are fasciculate (clustered or aggregated) (e.g. *A. erinacea*). Phyllodinous species are usually heteroblastic (Fig. 9.2), possessing compound leaves as seedlings that are soon replaced by phyllodes as the plant grows.
2. *Daviesia* (Fabaceae: Faboideae) – Phyllodes in this genus are either well developed, often spinescent, or are reduced to scales. When reduced to scales, branchlets (cladodes) then take over the photosynthetic function. Cladodes often end in a rigid, sharp point. Some species (e.g. *D. horrida*) produce phyllodes and cladodes on the same plant.
3. *Jacksonia* (Fabaceae: Faboideae) – Photosynthetic branchlets are flattened, angular or terete (= round in cross section) and often spinescent. Whereas adult leaves are reduced to scales, juvenile leaves are laminate and unifoliate (= a compound leaf appearing to be simple) when present. Other fabaceous genera (e.g. *Bossiaea*, *Templetonia*) also have species with reduced leaves and photosynthetic stems.
4. *Allocasuarina* and *Casuarina* (Casuarinaceae) – Outermost branchlets are the principle photosynthesising structure in this genus with the leaves reduced to whorled membranous scales. The green branchlets are sometimes called a cladode. Further, stomates are confined to grooves running along the stem (Chapter 3). The grooves close over as the stem dehydrates during drought, further restricting water loss.
5. *Exocarpos* and *Leptomeria* (Santalaceae) – Both these genera are root hemiparasites, possessing cladodes and leaves reduced to scales (in *Leptomeria* confined to species in Section *Leptomeria* of the genus).

Fig. 9.5: Selection of leaf-like structures (*from top to bottom*): vertically-oriented 2-cm-long phyllodes (true leaves have to twist vertical whereas phyllodes grow vertically) of *Acacia merrallii* (Fabaceae); pungent phyllodes of *Daviesia nudiflora* (Fabaceae) 3 cm long and *D. incrassata* 2 cm long; *D. euphorbioides* has prominent photosynthetic cladodes, 7 mm wide, with phyllodes reduced to recurved spines; yellow-flowered *Jacksonia sternbergiana* (Fabaceae) with thorny cladodes 5 cm long; the genus *Allocasuarina* (Casuarinaceae) is characterised by cladodinous branchlets with leaves reduced to scales and occurring in whorls around the nodes, pictured here is *A. humilis* with 5-cm-long branchlets; the bird-pollinated *Daviesia epiphyllum* (Fabaceae) with its 40-mm-long flowers seemingly arising from its leaves (hence the specific name) that in fact are flattened cladodes with phyllodes, or phylloclades. Cladodes are also a common feature in the root parasitic genus *Exocarpos* (Santalaceae), pictured here is *E. aphyllus*, with 5–10 long branchlets.

9.5 Alternatives to Sclerophylly for Protecting Leaves

Because the deposition of lignin and cellulose are time-dependent processes, young leaves are particularly vulnerable to herbivores, desiccation and scalding, especially as they are most exposed at the tips of the branches. Young leaves in many SouthWest evergreens are red due to the production of xanthophylls before the more nutrient-demanding synthesis of chlorophylls (Fig. 9.6). Since red reradiates in the 'hot' part of the light spectrum this can be expected to reduce the heat load on the leaf but experimental support is lacking. In contrast to mature leaves and leaves of adult plants that are more sclerophyllous and spinescent, young leaves and those on seedlings produce abundant phenolics that are effective against invertebrate herbivores (Hanley & Lamont, 2002; Hanley *et al.*, 2007). The young foliage of many species is covered in a dense indumentum of white or orange hairs (trichomes) that gradually abrade over time as the cuticle becomes much thicker. Other species retain their hair cover indefinitely giving a white, hoary appearance to the plant (Fig. 9.6). This pubescence is multifunctional, including increasing the saturated boundary layer to inhibit water loss, reflecting light and hence reducing the heat load, decreasing palatability and digestibility for herbivores, and deterring insects from ovipositing (Hanley *et al.*, 2007). Many evergreens have a glaucous appearance due to the secretion of surface waxes (Fig. 9.6) that appear to function in a similar way to trichomes.

Glandular hairs that secrete resin (terpenoid/flavonoid mix) are well represented in SouthWest species of the families Boraginaceae, Fabaceae (subfamily Mimosoideae), Goodeniaceae, Lamiaceae, Scrophulariaceae, Sapindaceae and Solanaceae (Dell, 1977; Fig. 9.7). They are probably universal in *Eremophila* (Scrophulariaceae), a large genus lacking scleromorphy (malacophyllous) and well represented in the heavier soils of the driest parts of the SouthWest. These hairs need to be distinguished from the mucilage-secreting hairs of carnivorous plants (Chapter 4). Resin-secreting hairs are usually most abundant on leaves but may also occur on dormant buds (e.g. *Banksia attenuata*, though resin glands have yet to be observed), young stems (e.g. *Grevillea eriostachya*), inflorescences (e.g. *Grevillea leucopteris*) and the lower parts of flowers (e.g. *Calytrix glutinosa*, Myrtaceae). These resins are viscous at first and flow over the surface and may contribute 15–20% of the dry weight of mature leaves (Dell & McComb, 1974, 1975). They often harden to form a varnish over the surface making it shiny and greatly increasing reflectance of all parts of the visible spectrum (Dell, 1977) and thus reducing heat load on the leaf. They are hydrophobic and must also act in a similar way to cutin in reducing water loss *via* the epidermis. Glandular hairs associated with flowers are sparser and the resin remains viscous without forming a sheet. They are more likely to have an antiherbivore function here by deterring and capturing insects in general and reducing palatability. Many *Stylidium* species possess glandular hairs on their stems and parts of inflorescences and flowers not directly related to reproduction but they appear to be mucilaginous rather than resinous (Darnowski *et al.*, 2006). They are rarely seen to capture insects and light reflection is not obvious so their function is unclear.

Fig. 9.6: Leaf surface features. (*upper left*) Young foliage of *Adenanthos cuneatus* (Proteaceae) is red and covered in a fine indumentum of hairs that is still present on mature leaves, 3 cm long, giving them a hoary appearance. Note their vertical orientation. (*upper right*) The entire shoot system (apart from the oldest basal stems) of *Eucalyptus macrocarpa* (Myrtaceae) is covered by a waxy glaucous bloom that is densest on the youngest leaves. Apical bud 2 cm wide. (*middle left*). *Hemiphora elderi* (Lamiaceae), 50 cm tall, not only has a dense indumentum over its leaves giving the plant a hoary appearance but also woolly flowers covered with deep-red, stellate hairs. (*middle right*) *Verreauxia reinwardtii* (Goodeniaceae) has 2-mm-thick, white wool over the entire plant. When examined under the electron microscope, the intricate structure of its stellate hairs are revealed with the tips of opposing branches fused by 'plugs' and short glandular hairs near the epidermis. Scale = 100 µm. (*lower left*) *Acacia denticulosa* (Fabaceae) showing a vertically-oriented phyllode, 6 cm long, covered in resin exuded by tuberculate glands over both (isobilateral) surfaces and giving it a highly reflective appearance. (*lower right*) Surface of young leaves, inflorescences and stems of *Grevillea leucopteris* (Proteaceae) are covered in both T-shaped hairs (a common feature of hakeas and grevilleas) and simple glandular hairs, 50 µm tall, that secrete resin. These hairs capture a wide variety of insects but whether they have any role in deterring herbivores, such as lepidopteran larvae, is unknown. With respect to possible carnivory, the average capture of 50 insects per 100 cm of branch length would only meet 0.15% of the nitrogen requirements of the plant even if fully released (Lamont, 1982c).

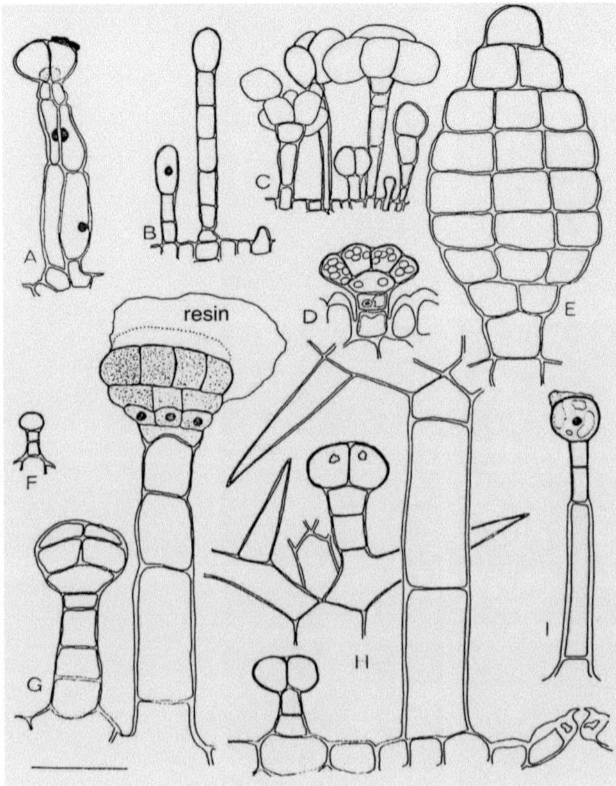

Fig. 9.7: Line drawings of glandular hairs (some bearing globs of resin) on leaves and stems with shiny or sticky surfaces due to the resin they produce flowing over the epidermis and reducing water loss and the summer heat load. A. *Helichrysum rosea* (Asteraceae), B. *Grevillea eriostachya* (Proteaceae), C. *Acacia aneura* (Fabaceae), D. *A. glutinosissima*, E. *Anthocercis littorea* (Solanaceae), F. *Scaevola crassinervia* (Goodeniaceae), G. *Scaevola glandulifera*, H. *Eremophila leucophylla* (Scrophulariaceae) (note stellate hairs may also be glandular and presence of raised surface stomates, usually a feature of mesophytes, that prevents resin flowing into pore and clogging it), I. *Pityrodia bartlingii* (Lamiaceae). Scale = 50 µm. From Dell (1977) and used with permission from the Royal Society of Western Australia.

9.6 Leaves as Defensive Mechanisms Against Herbivores, Florivores and Granivores

Most leaf herbivores in the SouthWest are insects, marsupials (e.g. possums, kangaroos), emus and introduced fauna (e.g. mainly rabbits). Leaves can display a number of structural defences to deter or reduce herbivory, including sharply-pointed spines and hardened leaf margins, hairs and tough (sclerophyllous) leaves (Fig. 9.8) (Hanley *et al.*, 2007). Within a species, seedling leaves often possess chemical deterrents to herbivores while mature plants are more likely to rely on physical deterrents, such as high LMA and spines (Rafferty *et al.*, 2005; 2010).

The production of distasteful or toxic compounds (e.g. phenolics, alkaloids, essential oils) as chemical defences is an alternative means of deterring herbivores. This may occur as a trade-off with structural traits, as for example within the genus *Gastrolobium* (Fabaceae) (Twigg & Socha, 1996). Highly spinescent and sclerophyllous *Gastrolobium* species (Fig. 9.8) tend to have lower leaf concentrations of the toxic chemical, sodium fluoroacetate, than less spiny, softer leaved species. Because all native vertebrate herbivores have evolved a natural tolerance to fluoroacetate, the commercial preparation known as 10-80 can be used in baiting programs to control feral cats and foxes.

Fig. 9.8: Sclerophyllous and spinsecent leaves of *Gastrolobium spinosum* (Fabaceae), leaves 4 cm long. (*upper middle*) *Allocasuarina pinaster* (Casuarinaceae), pungent cladodes 5 cm long. (*upper right*) *Grevillea hookeriana* subsp. *hookeriana* (Proteaceae), pronged terete leaves 3 cm long. (*lower left*) *Banksia heliantha* (= *Dryandra quercifolia*) with serrated leaves, flower head 10 cm diameter. (*lower middle*) *Astroloma glaucescens* (Ericaceae), leaves 2 cm long. (*lower right*) *Daviesia pachyphylla* (Fabaceae), commonly known as the ouch bush, because of the relatively long and sharply-pointed mucro at the tip of the phyllode (2.5 cm long).

Anti-herbivory traits may also be effective against florivores (flower eaters). Work on the genus *Hakea* (Proteaceae) (Hanley *et al.*, 2009) shows that insect-pollinated species (e.g. *H. costata*) (Fig. 9.9) tend to have small-flowered inflorescences either subtended by sharply pointed leaves or surrounded by dense sclerophyllous foliage. In contrast, bird-pollinated species (e.g. *H. francisiana*, Fig. 9.9) have inflorescences that are highly accessible, usually red in colour (See Chapter 8), and contain higher levels of floral cyanoglycosides (cyanide-based compounds) that deter florivores. The red floral coloration not only attracts visually cued pollinators (e.g. honeyeater birds) but is also a warning that the infloresecences are unpalatable.

Fig. 9.9: (*upper*) insect-pollinated *Hakea* species (from left to right) *H. costata* (with 15-mm-long *Stigmodera* (jewel) beetle pollinator), *H. spathulata* with 3-cm-wide axillary clusters and *H. gilbertii* with 2-cm-wide clusters. (*lower*) bird-pollinated *Hakea* species: *H. petiolaris* with 5-cm-wide clusters often produced on old stems (cauliflory), *H. cucullata* with 5-cm-wide cluster – the subtending leaves act as landing platforms and *H. francisiana* with 15-cm-long racemes. Note how the dense, spiny foliage of insect-pollinated species will deter avian florivores but bird-pollinated species have strong stems that also allow avian florivores, such as cockatoos, to land. However, these flowers contain high levels of toxic, cyanide-yielding compounds and there is a strong correlation between bright pink/red and the presence of cyanogens.

9.6.1 Cryptic Mimicry

Hakea trifurcata (Fig. 9.10) produces different types of leaves on the same branch (*i.e.* heterophylly): some are needle-like while others are broad and recurved. Instead of turning brown, mature woody fruits retain their green coloration. In turn, the broad leaves produced at the same axils as the fruits superficially resemble the fruits, their inrolling giving them a solid appearance. Studies have shown that this similarity confuses potential seed eaters, such as Carnaby's black cockatoo, and greatly reduces seed predation. Until they reach reproductive maturity, the juvenile plants only produce needle leaves. This unusual form of mimicry within a species is termed *self-crypsis* (Groom *et al.*, 1994). The broad leaves are often significantly larger than the fruits, and hence may also hide or shield the fruits from visually-cued granivorous birds.

Similar mimicry occurs within the *trifurcata* (*H. trifurcata, H. lasiantha, H. lasianthoides*) and *prostrata* (*Hakea prostrata, H. denticulata, H. pritzelii, H. amplexicaulis*) groups (Fig. 9.10; 9.11) all producing green fruits, sometimes with spiny or toothed appendages that mimic the surrounding leaves. The fruit walls are often the least woody among all hakeas, and the seeds are only retained in the crown for a year or so. The fruits of *H. stenocarpa* are brown, long, thin and gnarled and appear to resemble branchlets and the dead, strap-like leaves (that are retained). *H. ruscifolia* fruits are not only embedded among compact outward-facing leaves that have an extremely sharp mucro at the tip but they remain green and are a similar size and shape as the leaves.

9.6.2 Fruits Hidden by Leaves

Hakea polyanthema (Fig. 9.10; 9.11) has its woody fruits positioned deep within a 'nest' of outward projecting, sharply pointed, needle leaves, and creates an impregnable barrier, ensuring that the fruits remain inaccessible to potential granivores or insects attempting to lay eggs on the fruit surface. This is a feature of many spinescent hakeas in the SouthWest (e.g. *Hakea auriculata*; Fig. 9.10).

The cup or shell-like leaves of *Hakea conchifolia* and *H. smilacifolia* and undulating stiff leaves of *H. victoria* hide the fruits from granivores (Fig. 9.10; 9.11), decreasing the likelihood of seed loss, either from strong-billed cockatoos or seed-eating insect larvae.

Fig. 9.10: A selection of leaf-based strategies that minimise granivory in *Hakea*. (*upper left*) A heterophyllous branchlet of *H. trifurcata*. Arrows point to the woody fruits that remain green at maturity. Their shape and size resemble those of the surrounding broad leaves. Fruit 18 mm long. (*upper right*) Developing fruit (*centre*) of *H. denticulata* mimics the surrounding leaves in terms of coloration and marginal spines. Fruit 25 mm long. (*upper middle left*) *H. auriculata* with its 25-mm-long fruits enmeshed among extremely spinescent leaves. The unique prickles and brown markings on the fruits might have camouflage significance as well. (*upper middle right*) The mottled fruits of *H. polyanthema* embedded in an impenetrable nest of sharply-pointed, terete leaves. Fruits 25 mm long. (*lower middle left*) Mottled *H. erinacea* fruit partially camouflaged among its sharply-pointed, highly-divided, terete leaves. Fruit 20 mm long. (*lower middle right*) *H. stenocarpa* with four gnarled, elongated, 4-cm-long fruits that might be mistaken for branchlets or dead leaves by potential granivores. (*lower left*) Cup-shaped leaves of *H. smilacifolia* surround and conceal their small, thin-walled fruits from avian granivores. Leaf 30 mm long. (*lower right*) Woody fruits of *H. victoria* are well concealed by their subtending large, sharply-margined leaves. Leaves can be 15 cm wide. Photograph taken at an angle to the plant so that fruits are visible.

Fig. 9.11: Using a mummified Carnaby's black cockatoo and a recreated head using a skull and blue moulding clay to demonstrate how the spinescent leaves of selected *Hakea* species may deter avian granivores from attacking the woody fruits. Species are (*upper left*) *H. polyanthema*. Note that the pointed leaves are pressing against the bird's eye and beak. (*upper right*) *H. psilorrhyncha*. Despite possessing sharply-pointed, stiff leaves, the fruits of this species are sometimes easily accessable to cockatoos but they are especially large and woody. (*lower left*) *H. prostrata*. Leaves have sharp marginal spines, plus the fruit (pictured at the tip of the beak) resembles the surrounding leaves. (*lower right*) *H. auriculata*. This species is heterophyllous with the leaves associated with flowering (protecting the inflorescences as well as the fruits) are narrower and three-pronged at the apex, compared with the leaves not associated with flowering.

9.6.3 Herbivore Defence Among Restios

The Restionaceae (wire rushes) is a major herbaceous family in the SouthWest flora. Species have wiry green stems (culms) with leaves reduced to dry scales at the nodes. These are a poor food source for herbivores, averaging 0.57% N and 0.02% P, and high in fibrous tissues and tannins (Meney & Pate, 1999). Nevertheless, any plant that is grass-like in form is favoured by kangaroos and wallabies, and restios are often heavily browsed. However, 16 species of *Lepyrodia* and five species of *Loxocarya* produce a rare sulphur-containing amino acid, S-methyl cysteine, that strongly inhibits browsing by kangaroos and grasshoppers (Pate *et al.*, 1995).

10 Seed Release and Dispersal Mechanisms

For seedling recruitment to occur seeds need to be dispersed into an environment that promotes germination and seedling survival. Dispersal consists of two phases. Primary dispersal is defined as the initial transport of seeds or seed-bearing fruits (collectively seeds and fruits are called diaspores) to the ground or water body, or for aerial parasites, a host branch. Secondary dispersal relates to any subsequent movement to the seed's final resting place.

Primary dispersal may be active (e.g. seeds released explosively from the fruit, e.g. dehiscence (opening) of *Hardenbergia* pods), passive (e.g. seeds fall out when the capsules of *Eucalyptus* open), or require a vector to aid in seed removal (e.g. wind uplift of winged seeds of *Hakea* or winged fruits of *Nuytsia*; *Amyema* berries consumed by mistletoe birds).

Secondary dispersal involves either a biotic (e.g. ants) or environmental (e.g. wind, water) vector, and it is usually a different mechanism than that involved in primary dispersal. While primary dispersal is usually only for a few metres, secondary dispersal may cover several kilometres, and sometimes thousands for tiny seeds.

This chapter covers some of the dispersal mechanisms exhibited by the SouthWest flora following their release. Terminology used to describe seed dispersal mechanisms is provided in Table 10.1.

Table 10.1: Seed dispersal terminology.

Term	Definition
Anemochory	Wind dispersed
Chamaechory	Dispersal by rolling along the ground (wind assisted)
Zoochory	Animal dispersed (general)
Myrmecochory	Ant dispersed
Ornithochory	Bird dispersed
Mammalochory	Mammal dispersed
Hydrochory	Water dispersed
Barochory	Unassisted (gravity causes seeds to drop to the ground)
Autochory	Dispersal assisted by the actions of the parent plant (e.g. swaying in the wind)
Bolochory	Dispersal by propulsive mechanisms (ballistic)
Endozoochory	Animal dispersed, seed (or diaspore) eaten
Epizoochory	Animal dispersed; seed carried in fur, feathers, feet
Synzoochory	Animal dispersed, seeds carried intentionally (for consumption)
Dyszoochory	Animal dispersed, seed eaten intentionally
Nautohydrochory	Water dispersed, seeds carried in flowing water
Xerochasy	Seed release induced by fruit desiccation
Hydrochasy	Seed release induced by moisture

10.1 Pyriscence of Serotinous Fruiting Structures

Prior to their dispersal by wind, seeds must be released from the fruits or cones of serotinous species. This is usually through the action of fire, called pyriscence (Lamont, 1991). For other groups of plants, seed and fruit release is simply a matter of spontaneous abscission following maturation, often hastened by strong winds that dislodge the diaspores, as in *Nuytsia floribunda* (Chapter 5). Omnivores like emus pluck fruits from the plant. Seeds of bird-dispersed acacias, with bright-red arils, remain displayed in the open pods. Succulent diaspores, such as *Persoonia* fruits and *Macrozamia* seeds, drop to the ground when ripe where they are found by frugivores. Fruits of *Adenanthos cygnorum* are released into the surrounding leafy cup where they are located by granivorous ants (Fig. 10.6).

We illustrate the requirements for pyriscent seed release using the genus *Banksia* as it is by far the most interesting and best known. That fire is the key is supported by observations that the level of serotiny among three widespread banksias increases over a 500-km gradient north of Perth (Cowling & Lamont, 1985b). As the climate becomes drier, hotter and more seasonal so the height of the vegetation declines (scrub-heath) and fire is more likely to reach the position of the cones (crown rather than ground fires) so that serotiny can be relied on as a device for ensuring follicle opening and seed release at the time when seedling recruitment is most likely. In the open woodland further south, crown-reaching fires are less guaranteed and serotiny is weaker—interfire recruitment is more likely as well since the climate is milder. Most *Banksia* species retain their seeds in closed follicles for 5–10+ years accompanied by persistent dead florets—both traits are ancestral among banksias and can be traced back to the origin of the genus, 61 Ma (Fig. 10.1). Many species, such as *B. brownii*, never open all their follicles in the absence of fire, even when the plant dies from drought or disease (Lamont, 1996). Later (26 Ma), the persistence trait evolved further to conceal the fruits, possibly to hide them from efficient granivores like cockatoos whose ancestors were evolving at this time (White *et al.*, 2011) (Fig. 10.1).

Also at this time, species began to retain their dead foliage, a trait restricted to many SouthWest species and genera (Fig. 10.1). Both dead floret and leaf retention increase flammability (Lamont & Cowling, 1984) and, together with their narrow, sclerophyllous leaves, the plants burn strongly and quickly, ensuring that the resin sealing the valves of the follicles melts (Enright & Lamont, 1989; Fig. 10.2) and they open as they dry out. If the fire is intense enough, seeds may be released a few hours after fire. Otherwise, the onset of autumn rains results in wet-dry cycles that cause the gap to widen and the separator, which splits down the centre and clasps the wing of each seed in the illustrated *B. lanata*, gradually levers the two seeds out of the follicle (Cowling & Lamont, 1985a; Lamont & Barker, 1988). Seeds drop to the ground or become airborne at the level of the cone and are usually redistributed by wind and water into post-fire, litter microsites (Lamont *et al.*, 1993). Gusts of wind or wind vortices ('willy-willies'), particularly prominent after fire, may pick up the seeds and carry them several km from the parent plants (He *et al.* 2004; 2009). Black cockatoos may also remove the cones, especially after fire (note that the separator conceals and protects the seeds and makes it difficult for birds to access them while they remain on the plant), but the significance of this is unknown.

Fig. 10.1: Sequence of steps involved in the release of seeds from serotinous cones of *Banksia* (data refer to 85 banksias and 5 dryandras from He *et al.*, 2011). Pictured are: *B. lemanniana* (dead floret retention) - note that the cone hangs upside down in this species (it is wasp-pollinated) and that the first florets to open are the ones most likely to set fruits; *B. hookeriana* (dead florets conceal follicles); *B. candolleana* (dead leaf retention) a clonal shrub estimated to survive for up to 1,200 years (Merwin *et al.*, 2012), that also retains its dead florets on cauliflorous heads, ideally located for pollination by honey possums; *B. lanata* (resin melts).

Fig. 10.2: Removal of *Banksia attenuata* seeds following cone harvesting can be achieved by flaming the woody fruits. This melts the resin that holds the two valves together, causing the follicle to rupture immediately. Dunking the cone in water and allowing it to dry, and repeating many times, enables the follicle to open progressively and the enclosed separator gradually lifts the two seeds out of the fruit (Lamont & Barker, 1988). Cone is 15 cm long. Image provided by Haylee D'Agui.

10.2 Wind Dispersal

Seeds designed specifically for wind dispersal (anemochory) have structures that increase the diaspore's wind resistance and make it buoyant. Anemochorous seeds can be classified as either gliders, parachutes, helicopters, flutterers or tumbleweeds.

The three-winged, single-seeded fruit of *Nuytsia floribunda* (Loranthaceae) (Fig. 9.1) relies on strong gusts of wind in autumn to become airborne (Lamont, 1985b). Each fruit is held in place by three persistent bracteoles, preventing immediate dispersal. These winged diaspores can be dispersed at least 50 m from the parent plant, although the greatest density of fruits occurs within 10 m. *Nuytsia* fruits can be regarded as 'flutterers', with the wings enabling an increase in uplift by decreasing the fruit's terminal velocity, defined as the speed at which a seed ceases to accelerate because the downward force of gravity equals the upward force of air drag.

Seeds of *Banksia, Hakea* and *Xylomelum* (all Proteaceae) (Fig. 10.3) and *Allocasuarina* (Casuarinaceae) have diaspores subtended or encircled by a papery wing that accounts for at least half the diapore area. Casuarinaceae has woody bracts around what appears to be a winged seed but is actually a fruit. The fruits or woody cones open as a result of desiccation that is caused by death of the supporting stem from heat, drought or physical damage. Seeds can be released within a few hours of their protective woody-fruit opening in response to severe fire. With less intense fire and only partial opening of the fruit or cone, a wet-dry cycle is required (Cowling & Lamont, 1985a; Lamont, 1988).

With each successive bout of autumn rain and subsequent drying the woody structure opens more until finally the seeds are released onto an ideal seedbed for germination. This also minimizes the time of exposure to summer heat and granivores before germination can commence.

Taxa in the family Asteraceae (daisy family) characteristically produce a one-seeded fruit (achene) terminated by a feather-like crown of bristles (pappus). These act like a parachute—catching the wind so that the fruit floats in the air. Feathery parachutes also occur in other families (e.g. *Clematis*, Ranunculaceae). Hairs aid wind dispersal within certain Proteaceae (e.g. *Conospermum, Stirlingia, Petrophile, Isopogon*) and the awns of *Calytrix* (Myrtaceae) fruits, originally the calyx, have a similar function (Fig. 10.3).

Some species utilise prevailing winds to blow seed-laden fruits across the ground, dispersing seeds as they roll. *Spinifex longifolius* (Poaceae) fruits are aggregated into a spherical-shaped head that detaches and tumbles along the coastal sand dunes. This is assisted by the elongated bristles that subtend the individual fruits and are aligned radially.

Fig. 10.3: (*upper left*) Three-winged, single-seeded, fruit (samara) of *Nuytsia floribunda*. The surrounding woody bracts ensure that the fruits are only dispersed by strong gusts of wind. Winged fruits 2.5 cm wide. (*upper right*) Samaras of *Allocasuarina fraseriana* are protected by woody bracts. As the bracts dry the wing of the samara is exposed, in preparation for dispersal by wind. Samara 6 mm long. (*middle left*) Seeds of *Hakea platysperma* (left of image) are completely surrounded by a paper-thin wing. Each woody fruit (centre) contains two winged seeds. Seed plus wing 5 cm in diameter. (*middle right*) Open woody fruit (top) and winged seed (below) of *Xylomelum angustifolium*. Note the two empty seed cavities surrounded by the protective woody fruit walls. The seed proper (left) and wing (right) are both mottled, apparently providing camouflage from granivores in the post-fire environment. (*lower*) The awned calyx of *Calytrix* (Myrtaceae) flowers (*left*) is unusual in that it is persistent in the fruit (*right*) and aids in dispersal by wind.

10.2.1 Secondary Dispersal

Primary dispersal of woody-fruited Proteaceae occurs when the seeds or fruits are blown out of an open fruit or cone. When released, the seeds typically spin in a 'helicopter-like' motion, except for the very large-seeded *Hakea platysperma* that gracefully floats to the soil surface. Seeds may later be carried by wind or water across the ground (secondary dispersal) until they are permanently trapped by leaf litter, fallen twigs or other obstacles, or in soil depressions, or until the membranous wing is dislodged. The duration of wind-assisted seed movement is dependent on when a seed becomes entrapped by surface obstacles or covered by soil, or when the first heavy rains occur.

There has been much interest in long-distance dispersal among banksias. By genotyping individual plants within scattered populations across a dune system in the Eneabba Plain it has been shown that 2–6% of the plants in populations of three *Banksia* species must have migrated there (as seeds) from surrounding populations on average 1 km away but up to a distance of 3 km (He *et al.*, 2009; 2011). Clonal species such as *B. candolleana* have the lowest rates of secondary dispersal while fire-killed species such as *B. hookeriana* have the highest rates (Merwin, He, Lamont *et al.*, 2012). It takes on average 100 seeds to produce one adult plant of *B. hookeriana* compared with 4,500 seeds to produce one adult plant of *B. candolleana*. The inefficiency of recruitment among *B. candolleana* is countered by the fact that the longevity of this species has been estimated to reach 1,200 years, passing through scores of fire cycles in that time. In contrast, *B. hookeriana* survives up to 45 years but is usually killed within 20 years due to recurrent fires (Enright *et al.*, 1998).

The mechanism of long-distance transport among banksias is unknown but is most likely to be wind vortices that are common after fire when the seeds are released from the fruits (He *et al.*, 2004; 2009). The vortices probably pick up the seeds from the open cones or ground and carry them to a height of several kilometres when they drift back to earth in the direction of the prevailing winds. Since post-fire cones are often damaged by black cockatoos, as the seeds are vulnerable in the open fruits, it is possible that these granivores contribute to long-distance dispersal as well (He *et al.*, 2004). Damaged *Banksia prionotes* cones, for example, have been observed 250 m from the closest possible parent, and pine cones have been dropped by black cockatoos over 1 km from the nearest plantation.

10.3 Ant Dispersal

The SouthWest flora has the highest number of ant-dispersed taxa in the world (Lengyel *et al.* 2009; Mucina & Majer, 2012; Table 10.2), with dispersal generally dominated by *Rhytidoponera* (Formicidae: Ectatomminae) and to a lesser extent *Melophorus* (Formicinae) (Gove *et al.*, 2007; Majer *et al.*, 2009; 2010). The majority of ant-dispersed (myrmecochorous) plants inhabit the nutrient-impoverished sandplains

(Lengyel *et al.*, 2010), where more than a third of the shrubland sandplain flora is ant-dispersed. Myrmecochorous seeds bear a lipid and protein-rich appendage or aril, termed the elaiosome, that serves as a food reward for the ants, with ants known to respond to the chemical signals that they contain (Turner & Frederickson, 2013).

Table 10.2: List of SouthWest plant families and genera that are recorded as possing ant-dispersed species. Boldface type indicates genera with more than 50% of their species displaying myrmecochory. Data from Lengyel *et al.* (2009).

Aizoaceae	***Gunniopsis***
Amaranthaceae	*Sclerolaena, Maireana*
Apiaceae	***Platysace, Xanthosia, Actinotus***
Celastraceae	*Psammomoya*
Colchicaceae	***Burchardia***
Cyperaceae	*Lepidosperma, Carex*
Dasypogonaceae	***Dasypogon***
Dilleniaceae	***Hibbertia***
Elaeocarpaceae	***Tetratheca***
Ericaceae	*Leucopogon, Brachyloma, Monotoca*
Euphorbiaceae	***Bertya, Beyeria, Ricinocarpos, Euphorbia, Adriana, Monotaxis***
Fabaceae (Faboideae)	***Daviesia, Bossiaea, Goodia, Hardenbergia, Kennedia, Hovea, Templetonia (and others)***
Fabaceae (Mimosoideae)	***Acacia***
Goodeniaceae	***Dampiera, Goodenia, Scaevola, Velleia, Coopernookia***
Gyrostemonaceae	***Gyrostemon, Codonocarpus, Walteranthus***
Hemerocallidaceae	***Caesia, Hensmania, Johnsonia***
Proteaceae	***Grevillea, Adenanthos***
Rhamnaceae	***Pomaderris, Spyridium, Trymalium, Siegfriedia, Cryptandra***
Rubiaceae	***Opercularia, Pomax***
Rutaceae	***Asterolasia, Phebalium, Microcybe***
Sapindaceae	***Dodonaea***
Urticaceae	***Parietaria***
Violaceae	***Hybanthus***
Zygophyllaceae	***Zygophyllum***

Rhytidoponera violacea ants (Fig. 10.4) are keystone seed disperser in these shrublands (Lubertazzi *et al.*, 2010), storing seeds in nutrient-enriched middens inside their nests. Seeds can be re-dispersed from the nest by worker ants after the elaiosome hase been removed, often within 12 hours of being collected (Lubertazzi *et al.*, 2010). Elaiosomes are typically pale-coloured or white, smooth and smaller than the seed or indehiscent fruit to which it is attached (Fig. 10.5, 10.6). In some genera (e.g. *Xanthosia*, Apiaceae) it is hairy, while the brittle wing is attractive to ants is some grevilleas (Majer & Lamont, 1985)—more advanced grevilleas have a ridge of soft tissue around the seed more typical of other elaiosomes.

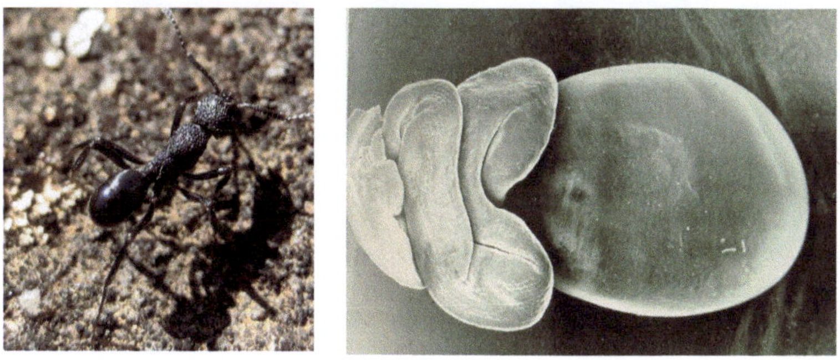

Fig. 10.4: Seed removal by ants is almost entirely due to two *Rhytidoponera* species, especially *R. violacea (left)*, in the northern sandplains (Gove *et al.*, 2007). Ant is 9 mm in length. Rhytidoponeras are relatively large ants and may prefer to remove large seeds like those that occur in the Fabacaeae, such as *Acacia lasiocalyx* (*scanning electron micrograph shown right*) that bears a highly convoluted basal elaiosome. Seed 7 mm long.

Most seeds secondarily dispersed by ants are primarily dispersed either ballistically (*i.e.* by expulsion) or simply fall out of the fruit, with ants harvesting seeds from the ground. Ants harvesting seeds from the plant itself is uncommon, but is known to occur in *Adenanthos cygnorum* (Proteaceae) (Fig. 10.7). Ants cannot account for long-distance seed dispersal of any species, as the mean dispersal distance is only 2 m from the parent plant, so that a third dispersal agent, such as emus, may be required to carry them even further.

In the SouthWest, all ant taxa create underground nests, so that myrmecochorous seeds are removed from the soil surface and buried. Once carried underground, the elaiosome is consumed by the ants and the intact seed discarded as waste in an underground or surface midden. Elaiosome-lacking seeds discarded on the soil surface may potentially undergo further dispersal by other ant species. Ants may also move seeds that do not possess elaisosomes [e.g. *Darwinia* (Myrtaceae) (Auld, 2009) though this has a fleshy exocarp that is palatable] and not all elaiosome-bearing seeds are solely dispersed by ants [e.g. *Daviesia triflora* (Fabaceae) is also dispersed by emus up to 2.5 km from the parent plants (He *et al.*, 2009)]. Some Ericaceae produce fleshy fruits that are predominantly bird-dispersed, although the fruits may also attract ants (Keighery, 1980).

10.3.1 Fate of Ant-dispersed Seeds

After release to the ground surface, a seed adapted for ant dispersal has many possible fates (Fig. 10.5). Seeds are subjected to pressures from granivores. Being eaten by a vertebrate may aid in long-distance dispersal if the seed survives the digestive process. Once transported to the ant nest the depth of burial is crucial for the seed's long-term survival, as this has implications for dormancy breaking (receiving the fire heat pulse) and the ability of the developing seedling to reach the soil surface.

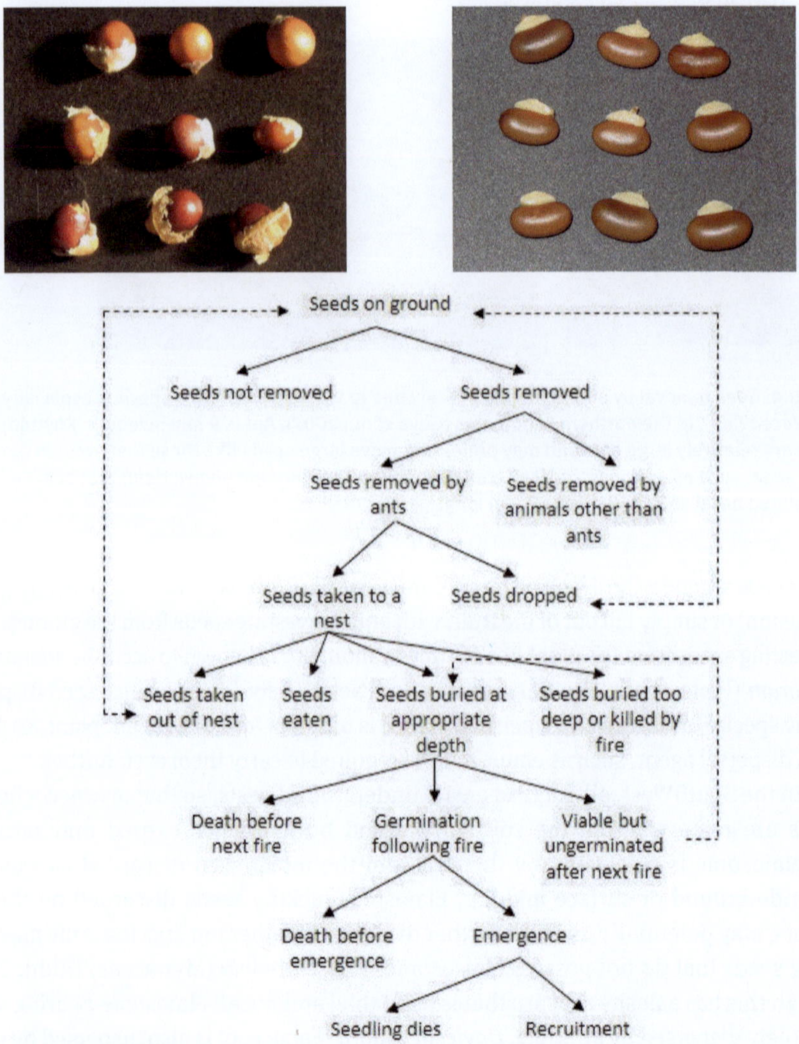

Fig. 10.5: (*upper*) Seeds of *Hibbertia hypericoides* (Dilleniaceae) (*left*) and *Hardenbergia comptoniana* (Fabaceae) (*right*) bearing elaiosomes. Seeds are 2 mm and 5 mm in length respectively. (*lower*) Potential fate of myrmecochorous seeds. Adapted from Hughes and Westoby (1992).

Seeds buried beneath the soil are protected from granivores and desiccation, and will not be consumed by a passing fire. However, many Australian ant-dispersed seeds require fire to break dormancy and promote germination. Most myrmecochorous seeds have hard testas preventing the seed from being consumed by granivorous ants (Fig. 10.5, 10.6). This hard seedcoat also renders the seed impervious to moisture. Heat from an above-ground fire is sufficient to crack the cuticle or dislodge the strophiolar plug (Dell, 1980) of seeds buried relatively close to the surface without killing them. This enables the seeds to imbibe water and germinate during the next winter rain event.

Fig. 10.6: (*upper*) *Stachystemon axillaris* (Euphorbiaceae) is a slender, monoecious shrub in the northern Swan and Eneabba sandplains (plant in centre of photo is 50 cm tall). It usually only occurs underneath clumps of *Eucalyptus todtiana* (tree in background). (*lower left*) Its 4-mm-long seeds have a pale green aril (elaiosome) that is attractive to at least six ant species. Granivorous birds, such as bronze-wing pigeons, also consume the seeds and disperse them between *E. todtiana* clumps but feeding trials have yet to be undertaken. (*lower right*) Seeds are buried by ants in seed middens in their nests. The heat of a fire stimulates them to germinate *en masse* provided that they are located within 1–3 cm of the surface. The tallest seedling here is 8 cm. Recruited plants survive for 30 years, and since interfire germination is negligible and the stored seeds gradually die, if there is no new fire within 45 years the population dies out. Periodic fire is an essential part of the life cycle of *Stachystemon*. Photos taken by Allan Tinker.

10.3.2 Case Study: *Adenanthos cygnorum*

The woolly bush, *Adenanthos cygnorum* (Proteaceae), is a major colonizer of the Swan Coastal Plain after soil disturbance or fire. Its drab cypress-like appearance belies its fascinating biology (Fig. 10.7). *Adenanthos* is the only genus in the Proteaceae that produces isolated flowers—these produce copious nectar in their 'gullet' that attracts small honeyeater birds (Meliphagidae) even though they are not conspicuous (Fig. 10.7). *Adenanthos* probably had its origins in the SouthWest and is basal to a clade that arose about 45 million years ago, which otherwise accounts for most of the Proteaceae in South Africa (Chapter 1). This is one of the oldest clades known to produce oil-rich bodies (elaiosomes) attached to its diaspores (in this case, single-seeded fruits) attractive to ants that bury them in their nests. For *A. cygnorum*, many ant species are attracted to the plant initially *via* nectar-secreting extrafloral nectaries at the tips of their leaf segments. These nectaries are particularly well developed on the first leaves produced in the spring-flush of new growth (Chapter 7).

Fig. 10.7: *(upper left)* Adult plant of *Adenanthos cygnorum* (Proteaceae), 2 m tall. *(upper right)* Solitary flower, 25 mm long, partly embedded in foliage. *(lower left)* *Iridomyrmex bicknelli* (Formicidae: Dolichoderinae) removing fruit, 7 mm long, from the leafy cup in which it is held after release. Note elaiosome beneath thorax of the ant. *(lower right)* *A. cygnorum* fruits with and lacking the hair-covered elaiosome.

Also attracted are parasitoid wasps, *Campoletis* sp. (Ichneumonidae), that also feed from the nectaries. While not usually attractive to herbivores, there is a specialist phytophagous moth, *Xylorycta* sp. (Xyloryctidae) that webs together the terminal leaves and destroys the stem apices. The wasp is so efficient at detecting larvae in the webs and ovipositing on them that 70% of xyloryctid pupae may actually hatch as *Campoletis* wasps (Lamont, 1989). Also, while foraging on the plant, the ants are likely to come across mature fruits in summer-autumn that have been released into the surrounding cup of leaves created by the bracteoles that push them outwards as they spread sideways while desiccating during seed release. All other elaiosome-bearing species release their seeds onto the ground where, in this case, they would be vulnerable to granivores, such as pigeons, ground parrots and rodents. Small ants may cooperate to pick up the seed and carry it down the plant and into their nest (Fig. 10.7). Interestingly, many ants build their nests close to patches of the woolly bush, so good is it as a nutrient source. This means that the ants serve more to store the seeds effectively rather than disperse them to long distances from the parent.

The seeds may lie dormant in the nests for some years until brought to the surface by soil disturbance, or a fire passes over the patch. *A. cygnorum* is highly flammable and killed by fire. Scores of seeds around the parent plants may germinate in response to the increased temperature fluctuations associated with litter and canopy removal and greater exposure of the soil surface to light. A dense stand of the woolly bush may develop again even when the previous patch has long since perished.

10.4 Vertebrate Dispersal

As a general rule, seeds with large, brightly-coloured appendages, or fleshy, non-green fruits are ingested and dispersed by vertebrates (endozoochory) and have the potential to be dispersed over relatively long distances (Fig. 10.8). In the SouthWest flora this tends to be undertaken by birds (ornithochory) and to a lesser extent by mammals (mammalochory). Bird-dispersed fruits and seeds are coloured red, yellow or black. Fruits and seeds eaten by mammals are usually not as colourful and may be brown, green, dark red or orange. Seed dispersal by reptiles (saurochory) is rare, apart from bobtail lizards. Some plants rely solely on a particular animal (e.g. *Amyema* and the mistletoe bird), whereas some animals (e.g. emus) consume the seeds and fruits from a wide range of families and genera (Table 10.3).

Genera that produce endozoochorous fruits include *Rhagodia* (Chenopodiaceae), *Eremophila* (Scrophulariaceae), *Persoonia* (Proteaceae), *Pittosporum* (Pittosporaceae), *Solanum* (Solanaceae), *Cassytha* (Lauraceae), *Olax* (Olaceae), *Santalum* and other members of the Santalaceae, and the cycad, *Macrozamia*. All of these have been observed in the droppings of emus (Table 10.3). Rodents have been observed to consume the fleshy outer layer (sarcotesta) of *Macrozamia riedlei* seeds, but no actual dispersal has been recorded (Burbidge & Whelan, 1982). Brush-tailed possums have been reported to carry seeds up to 24 m from a female macrozamia.

Fig. 10.8: (*upper left*) *Acacia cyclops* (Fabaceae). Seeds, 6 mm long, are surrounded by a coiled double-backed, orange-red aril that is attractive to granivorous birds. The hard-coated seeds are often retained in the pod when it splits open (making it easy for birds to locate them), whereas ant-dispersed seeds are explosively released from the pods. (*upper right*) *Pittosporum angustifolium* (Pittosporaceae). The open fruit, 15 mm long, reveals several sticky seeds that are brown or black when dry. The genus name refers to the sticky resin that surrounds the seeds. (*lower*) *Santalum acuminatum* (Santalaceae). The comparatively large (3 cm wide) fleshy and waxy skinned, red fruits are easily spotted among the surrounding foliage and are a favoured food source for the emu.

Within the Ericaeae, subfamily Epacridoideae, the fruits of 25 species are known to be consumed by six bird species. These species either have fleshy drupes (e.g. *Acrotriche, Astroloma, Brachyloma, and Leucopogon*) or dry drupes (e.g. *Conostephium, Coleanthera*, and *Styphelia*) and all are endozoochorous (Keighery, 1996).

10.4.1 Emus and large Macropods as Long-Distance Seed Dispersal Agents

The emu (*Dromaius novae-hollandiae*: *Dromaiidae*) is a large, flightless, omnivorous bird that is distributed over much of Australia, in a wide range of habitats, particularly semi-arid areas. This bird is known to feed upon a wide range of plant taxa (Table 10.3), though many are not primarily adapted for ornithochory. 40% of the average diaspore content of emu droppings are endozoochorous, and hence are adapted to survive ingestion and digestion as part of the dispersal process (Calviño-Cancela *et al.*, 2006). Barochorous species (no apparent structure or adaptation to facilitate dispersal) account for 30%, while the remainder is primarily wind or ant-dispersed. Not all seeds consumed by emus will survive the digestive process, and only some of the species with viable seeds that germinate in emu droppings actually require the digestive process to break seed dormancy. Emus play an important role in the long-distance transport of seeds over relatively short time periods.

The bright orange seeds of *Macrozamia* (Zamiaceae) (Fig. 10.9) are attractive to emus, rodents and bobtail lizards that consume the fleshy sarcotesta but only emus are known to act as long-distance dispersal agents. The seeds contain a poisonous cardioglycoside and oxalates in the endosperm that is also protected by a bony endotesta. Emus that swallow and defecate the entire seed, minus the fleshy layer, are not affected by the toxins.

Kangaroos, and other large macropods, also feed on a great variety of plant taxa, and are known to disperse seeds undamaged in their droppings, as well as lodged in their fur (Calviño-Cancela *et al.*, 2008). Kangaroos, and other herbivorous mammals, chew food before swallowing, thus only small, hard seeds remain undamaged (e.g. *Acacia blakelyi*). Both emus and kangaroos are capable of transporting seeds over several kilometres, and this enables certain plant taxa, e.g. *Daviesia triflora* (Fig. 10.9), to colonise other suitable habitats, potentially improving gene flow and connectivity between populations. For example, by using genotype assignment techniques, *Daviesia triflora* (Fig. 10.9) has been recorded up to 2.5 km from its parent population (Calviño-Cancela *et al.*, 2008; He *et al.*, 2009).

Fig. 10.9: (*left*) Seeds, 5 cm long, drop passively to the ground from the female cone of the cycad *Macrozamia riedlei* (Zamiaceae). A wide variety of animals feeds on the seeds but only emus distribute them widely over the landscape. (*right*) Flowers of *Daviesia triflora* (Fabaceae), 10 mm wide, whose seeds are released explosively from their pods then dispersed by ants (short distances) and, sometimes, by emus (long distances).

Table 10.3: List of SouthWest plant species located in emu droppings, either as seeds or fruits collected from the Eneabba sandplain, Fitzgerald River National Park and Cape Range National Park. Their usual mode of dispersal is indicated.

Family **Species**	**Mode of dispersal**	**Family** **Species**	**Mode of dispersal**
Amaranthaceae		Lauraceae	
Ptilotus spp.	Wind	*Cassytha* spp.	Endozoochory
Asparagaceae		Mimosoideae (Fabaceae)	
Thysanotus spp.	Ant	*Acacia blakelyi*	Ant
Boraginaceae		*Acacia cyclops*	Endozoochory
Trichodesma zelanicum	Unassisted	*Acacia farnesiana*	Unassisted
Faboideae (Fabaceae)		Myrtaceae	
Senna spp.	Unassisted	*Eucalyptus* spp.	Unassisted
Casuarinaceae		Olaceae	
Allocasuarina humilis	Wind	*Olax aurantia*	Endozoochory
Chenopodiaceae		Faboideae (Fabaceae)	
Maireana spp.	Wind	*Daviesia teretifolia*	Ant
Rhagodia eremaea	Endozoochory	*Swainsona pterostylis*	Unassisted
Dasypogonaceae		Portulacaceae	
Acanthocarpus preissii	Unassisted	*Calandrinia* spp.	Unassisted
Dilleniaceae		Proteaceae	
Hibbertia hypericoides	Ant	*Banksia shuttleworthiana*	Wind
Ecdeiocoleaceae		*Hakea ruscifolia*	Wind
Ecdeiocolea monostachya	Unassisted	*Isopogon trilobus*	Wind
Ericaceae		Santalaceae	
Astroloma microdonta	Endozoochory	*Exocarpos sparteus*	Endozoochory
Leucopogon spp.	Endozoochory	*Santalum acuminatum*	Endozoochory
Euphorbiaceae		*Santalum lanceolatum*	Endozoochory
Adriana tomentosa	Ant	Scrophulariaceae	
Goodeniaceae		*Eremophila* spp.	Endozoochory
Goodenia spp.	Wind	Solanaceae	
Gyrostemonaceae		*Solanum symonii*	Endozoochory
Codonocarpus cotinifolius	Ant	Surianaceae	
Tersonia cyathiflora	Endozoochory	*Stylobasium spathulatum*	Endozoochory
Haemodoraceae		Zamiaceae	
Haemodorum spicatum	Wind	*Macrozamia fraseri*	Endozoochory

Dispersal modes categorised are based on the following attributes—Unassisted: no apparent structures to facilitate dispersal; Ant-dispersed: elaiosomes present; Endozoochory: possess fleshy structures; wind-dispersed: possess structures that enhance air resistance. Data collated from Appendix 1 of Calviño-Cancela et al. (2006).

10.5 Unassisted Dispersal

Plant species that have no specialised means of dispersal have seeds that simply fall out of the open fruit, possibly dislodged by wind or swaying of fruits. Example genera are *Eucalyptus, Melaleuca, Calothamnus, Corymbia* (all Myrtaceae) and *Xanthorrhoea* (Xanthorrhoeaceae) (Fig. 10.10). Addition of the mature fruits may fall from the plant and the seeds can be released later (e.g. *Astartea*, Myrtaceae), or not at all (indehiscent). Seeds may be blown along the ground by gusts of wind post-release to reach their final resting place, but are not specifically adapted to do so.

Fig. 10.10: (*left*) *Corymbia calophylla* (Myrtaceae) seeds, 10 mm in length. (*right*) *Xanthorrhoea preissii* seeds resting in 10-mm-long, open fruits. Note the sharp tips to the valves that will deter avian granivores from accessing the seeds.

10.5.1 Kangaroo Paws

Most *Anigozanthos* species have small irregular-shaped seeds that are shaken out of the fruit by gusts of wind. Two species, *A. pulcherrimus* and *A. rufus* (Fig. 10.11), retain their seeds within the fruit. When the senescent fruit falls to the ground, seeds are eventually released as the fruit breaks up. For the black kangaroo paw (*Macropidia fuliginosa*) ripe seeds simply fall away from the fruit.

Fig. 10.11: *Anigozanthos* (Haemodoraceae) species in flower. (*left*) *A. rufus*, (*middle*) *A. manglesii* ssp. *manglesii* and (*right*) *A. preissii* that have no special traits for promoting dispersal other than their small size. All flowers 4–6 cm long. These are also good examples of bird-pollinated species with their terminal location, bright colours and reproductive parts facing towards the stem where the birds land. Images provided by Katherine Downes.

10.5.2 Orchid Seeds

Orchids produce numerous microscopic (< 0.1 mm diameter) seeds. One gram of seeds may contain more than 1 million seeds. Seeds are wind dispersed from a fruit that has 3 longitudinal slits that split open (capsule), releasing a cloud of seeds that lie dormant in the top soil until the next winter rains (Batty *et al.*, 2000; 2001). Orchid seeds are the smallest of all as they contain no endosperm and the embryo is incompletely formed (proembryonic). They can be dispersed by wind for hundreds of kilometres but have no special buoyancy mechanism to do so. They are obligately dependent on mycorrhizal fungi for seed germination and plant establishment (Rasmussen & Rasmussen, 2009; Swarts *et al.* 2010; Phillips *et al.*, 2011).

11 Seed Storage, Germination and Establishment

Seeds rarely germinate as soon as they are released because conditions for germination are unsuitable. Most are stored for a few months to many years before they can germinate. SouthWest species in 90 families are known to store their seeds until stimulated to germinate by fire (Enright *et al.*, 2007; Lamont & He, 2012). There are two types of seed storage: on the plant is called serotiny and occurs in 20-30% of the SouthWest sandplain flora (Lamont *et al.*, 1991; Lamont & Enright, 2000; Fig. 11.1). Persistent soil-stored seed banks are the more common seed storage syndrome exhibited by the SouthWest flora, and account for 65-75% of sandplain species (Enright *et al.*, 2007). For the SouthWest Proteaceae, where seed storage syndromes vary between genera, the differing ability to store seeds in the crown or the soil has implications for dispersal mechanisms and cues for germination (Table 11.1). The evolutionary stimulus for seed storage is that seeds produced in any year are inadequate for self-replacement should the parents die as is most likely after fire. Not only does progressive seed accumulation increase that available following disturbance, it also extends the fire interval tolerated at either end of the range (Lamont & Enright, 2000).

Fig. 11.1: (*left*) Spring flowering of low heath at Crystal Brook Reserve, 15 km east of Perth, on the shallow soils over Cenozoic laterite with ancient PreCambrian outcropping granite (on right). Over 80% of species visible in the photo [e.g. *Grevillea* (cream), *Hemigenia* (mauve), *Oxylobium* (orange), *Acacia* (yellow), *Diplopeltis* (pink), *Burchardia* (white 'spots')] recover after fire from soil-stored seeds. The remainder [*Hakea* (cream shrub in background), *Allocasuarina* (brown (male) and red (female) flowers), *Isopogon* (pink with yellow/red style), *Banksia* (*Dryandra*) (yellow)] recover from serotinous seeds. The one exception is *Xanthorrhoea preissii* (grasstree in mid-background with spent spikes produced in response to the fire 7 years before) that has fire-stimulated flowering with seeds released the year after fire that must germinate then or perish (Lamont *et al.*, 2004). (*right*) 1.5-m-tall stand of the bird-pollinated, fire-killed, serotinous *Banksia hookeriana* at Yardonogo Reserve. This species arose 5 Ma (He *et al.*, 2011). While the seeds stored by this species only weigh 0.5% of the total aboveground mass, they contain 48% of its phosphorus (P), 24% of its nitrogen (N) and 4% of its potassium (K). Pot trials have shown that the deep sands in which this species occurs are P, N and K-limited and growth of wild oats is negligible unless all three nutrients are added (Lamont, 1995).

Table 11.1: Seed storage syndromes of SouthWest Proteaceae. Adapted from Lamont & Groom (1998)

	Banksia, Hakea	*Isopogon, Petrophile*	*Conospermum, Persoonia*	*Adenanthos, Grevillea*
Storage	On plant	On plant	In/on soil	In soil
Fruit	Woody follicle	Hairy nut	Drupe, hairy nut	Achene, follicle (leathery)
Predispersal protection	Strong	Moderate	Nil	Weak
Seed release	Fire, rain, death	Fire, death	Spontaneous	Spontaneous
Diaspore type	Soft-coated, winged, SL	Hard-coated, SL	Hard-coated, LL	Hard-coated, LL
Dispersal agent	Wind	Wind	Frugivores, wind	Ants, wind
Germination stimulus	Winter rain	Winter rain	Smoke, wall decay	Smoke, light

SL = short lived; LL = long lived

The term serotiny refers to the retention of seeds within the crown of the plant for an extended time (hence canopy seed storage) (Lamont *et al.*, 1991). The degree of serotiny varies depending on the extent to which viable seeds are retained with increasing age of the fruit or cone. Non-serotinous species release their seeds as soon as they reach maturity. Species classified as weakly serotinous release most seeds within a year of seed maturity. If seeds are released within 5 years of seed maturity the species can be classified as moderately serotinous. Strongly serotinous species retain seeds in their crown for extended periods of time, often greater than 5 years and sometimes up to 15 years. Serotinous seeds are protected by woody fruits, either isolated, clustered or aggregated into an infructescence ('cone'), or with chitinous fruits protected by woody bracts in cones (Lamont *et al.*, 1991; Lamont & Enright, 2000; Fig. 11.1).

11.1 Canopy-stored Seeds

Retention of seeds in woody fruits or cones is a seed storage strategy (Fig. 11.2) and best represented in the SouthWest among all the world's floras. It is confined to a few dominant plant families and genera (Table 11.2). Most species that retain canopy-stored seeds release them *en masse* after the passage of fire (called pyriscence), as it is the heat from the fire that cues seed release. Thus, two distinctive adaptations are required: 1) a means for protecting the seeds from damage while retained on the plant, and 2) a means for releasing seeds in response to the heat of a fire.

Fig. 11.2: (*upper left*) A woody fruit, 5 cm long, of *Hakea neurophylla* (Proteaceae). Each fruit contains two winged seeds that are wind dispersed. The fruit remains closed on the plant until the above-ground stems are killed, usually by the heat of a fire, but sometimes through physical damage, disease or drought. (*upper right*) Woody cones, 4 cm long, of *Allocasuarina campestris* (Casuarinaceae). A pair of woody valves surrounds and protects each winged fruit. Valves remain closed for many years, opening as a result of desiccation, usually from branch death as a result of fire. (*middle left*) Woody cone of *Banksia lemanniana* (Proteaceae) with several prominent woody fruits (3 cm long) each protecting 2 winged seeds. *B. lemanniana* is one of the few banksias where the inflorescences hang downwards. (*middle right*) Clustered woody fruit of *Calothamnus graniticus* (Myrtaceae). Fruits are 2 cm long. (*lower left*) Cup-shaped fruits (8 mm long) of *Eucalyptus marginata* (Myrtaceae). (*lower right*) Fruit cluster of *Melaleuca thymoides* (Myrtaceae). Note the subterminal spinescent leafless branchlets (thorns) that presumably act to deter granivores. Cluster 4 cm long.

Some canopy-stored seeds may be released in the absence of fire, opening with age, or from exposure to sunlight, or as plants or branches die back or become damaged. Some species (e.g. the *Hakea prostrata* group) only retain their seeds for 1 year, but some banksias retain viable seeds for up to 15 years. Once seeds are released from their protective woody fruit or cone, they are susceptible to becoming dehydrated or scalded from summer heat or eaten by granivores. Seeds must germinate during their first winter or they will perish. Serotiny is more prominent on nutrient-impoverished soils because seeds are more nutrient-enriched (Groom & Lamont, 2010; Lamont & Groom, 2013) and thus subjected to increased post-dispersal granivory on the soil surface (Keeley *et al.*, 2011). However the fitness benefit of serotiny is undoubtedly related to cueing seed release into the post-fire environment, providing the best opportunity for germination and seedling recruitment to occur (Lamont & Enright, 2000).

Table 11.2: SouthWest plant groups that contain serotinous genera. Adapted from Lamont *et al.* (1991).

Family	Genus
Conifers	
Cupressaceae	*Callitris/Actinostrobus* (now placed in *Callitris*)
Angiosperms	
Casuarinaceae	*Allocasuarina*
	Casuarina
Myrtaceae, Leptospermoideae	*Agonis*
	Beaufortia
	Callistemon
	Calothamnus
	Conothamnus
	Corymbia
	Eremaea
	Eucalyptus
	Kunzea
	Leptospermum
	Melaleuca
	Regelia
Proteaceae, Grevilleoideae	*Banksia/Dryandra* (now placed in *Banksia*)
	Hakea
	Lambertia
	Petrophile
	Strangea
	Xylomelum
Proteaceae, Proteoideae	*Isopogon*

11.1.1 Adaptive Advantages of Canopy-stored Seeds

For the SouthWest flora, serotiny is a mechanism designed to maximise seed availability for release post-fire, ensuring that the simultaneous release of the seeds coincides with the presence of favourable conditions for germination and seedling recruitment (Lamont *et al.*, 1991; Enright *et al.*, 2007). Woody fruiting structures protect seeds from the heat of a fire (Bradstock *et al.*, 1994; Hanley & Lamont, 2000; Kim *et al.*, 2009), with the extent of seed release depending on fire intensity. Delayed release also allows seeds to accumulate more nutrients that increases their fitness in a nutrient-limited environment (Groom & Lamont, 2010). Weak serotiny is more common where conditions are suitable for recruitment even in the absence of fire, such as in the more coastal and wetter southern parts of the SouthWest, beside swamps and on rock outcrops. Seeds are more likely to germinate in the litter-filled microsites compared with areas of bare sand. However, the resulting seedlings are in greater competition with each other, especially for soil moisture, and are more likely to die (Lamont *et al.*, 1993).

Weakly serotinous species (e.g. *Hakea trifurcata*, *Banksia* (*Dryandra*) *sessilis*) release seeds between as well as after fires, and this can be considered a 'bet-hedging' strategy, creating the opportunity for limited inter-fire recruitment to occur in habitats that experience long periods without fire. So net recruitment is finally quite similar regardless of the degree of serotiny.

11.1.2 Protective Properties of Woody Fruits

Persistent canopy seedbanks are exposed to the pressures of predispersal granivory and desiccation. The most serotinous structures also are the largest and woodiest and thus they have the best-protected seeds (Groom & Lamont, 1997).

Woody fruits or cones provide thermal insulation from hot summer temperatures and fire-induced heat (Mercer *et al.*, 1994). They also give protection from, and a deterrent to, granivores, especially strong-billed birds and boring insect larvae (Fig. 11.4). Investing in woody fruiting structures that provide some protection for their seeds is a common adaptive response in fire-prone, nutrient-impoverished environments (Lamont *et al.*, 1991) and a special feature of the SouthWest flora. The degree to which seeds are retained within these structures has implications on the cost and ecological benefit of woody fruiting bodies (Lamont *et al.*, 1991; Midgley, 2000; Cramer & Midgley, 2009), and has a direct influence on seed maturation times, seed nutrient content and the degree of pre-dispersal protection.

Woody fruits are subjected to fire-induced heat over seconds, not minutes, (Judd, 1993; 1994) and are not designed for absolute heat protection. Clustering of fruits protects the sides and base of individual fruits (e.g. *Melaleuca*, Fig. 11.2) (Judd & Ashton, 1991; Brown & Whelan, 1998), and the most important heat protective features are the fruit's wood density, moisture content and radius (equivalent to wall thickness) (Mercer *et al.*, 1994). Increasing fruit radius from 3 to 5 mm reduces (halves) the maximum temperature reached within the fruit (Mercer *et al.*, 1994). Such small-sized fruits are typical of some *Eucalyptus* and other Myrtaceae with woody capsules (e.g. *Melaleuca*), but not of the much larger, woody-fruited Proteaceae.

Serotinous *Banksia* (Proteaceae) species proliferated as speciation accelerated (and/or extinction rates decreased significantly) during the cooler, drier more seasonal climate of the Miocene, with the proliferation of serotinous taxa slowing during the Pliocene and Quaternary (He *et al.*, 2011). In contrast to *Banksia*, *Hakea* (Proteaceae) is a more recent addition to the Australian flora, appearing 18 Ma during the mid-Miocene (Fig. 1.5), a period of increasing fire frequencies. *Hakea* is synonymous with serotinous, thick-walled, woody fruits, unlike its sister genus *Grevillea* that is non-serotinous. Woody fruits of SouthWest *Hakea* (Proteaceae) species (Fig. 11.3) vary in dry weight from 20 mg (*H. sulcata*) to 40,000 mg (*H. platysperma*) and vary in their ability to retain stored seeds (Groom & Lamont, 1996b). Groom and Lamont (1997) showed that strongly serotinous species (defined as seed retention time > 5 years) were more likely to have heavier, woodier and thicker-walled follicles than non- and weakly serotinous species (Fig. 11.3), and this should relate to increased seed protection because the thermal insulation properties of woody fruit walls are dependent on wall density and wall thickness (Bradstock *et al.*, 1994; Mercer *et al.*, 1994). A doubling in the radius of woody fruits results in a four fold increase in the exposure time necessary to cause heat-induced seed death (Mercer *et al.*, 1994).

Small, thin-walled *Hakea* fruits are poor insulators of heat (Bradstock *et al.*, 1994). However, because most seedling recruitment occurs after a fire, even small fruits must provide adequate heat protection for the seeds to survive. *Hakea* seeds are a major source of P for newly emerging seedlings (Groom & Lamont, 2010; Lamont & Groom, 2013) in the nutrient-impoverished soils of the SouthWest. N- and P-enriched seeds are a valuable food source for avian granivores and insect larvae, and may explain why species such as *H. platysperma* produce extraordinarily large woody fruits (Fig. 11.3). With walls 13–20 mm thick, *H. platysperma* fruits provide more than enough thermal protection for their seeds; thus the additional protection must be to minimise seed loss caused by the increased pressures of granivory.

Moderately/strongly serotinous *Hakea* species invest more energy (higher fruit:seed mass ratio) than non- and weakly serotinous species, consistent with increased seed protection. Generally, the greater the level of serotiny, the larger the seed and higher the total nutrient content, and greater the mass of the supporting fruit/cone (Groom & Lamont, 2010). However, the associated costs (*i.e.* total dry mass, inorganic nutrients) involved with producing woody fruits increases more sharply as *Hakea* seeds become larger than does investment in seed protection (*i.e.* follicle wall thickness) (Midgley *et al.*, 1991; Groom & Lamont, 1997). *Hakea* arose 18 Ma and its radiation corresponded with the presence of strong-billed, granivorous, black cockatoos (*Calyptorhynchus*: Calyptorhynchinae) (Fig. 11.4) that appeared with the expansion of Australia's sclerophyllous vegetation and eucalypt-dominated landscapes (White *et al.*, 2011). Black cockatoos specialize in extracting seeds from woody fruits (Cooper *et al.*, 2002; 2003) and therefore can be considered an important selective agent on fruit wall thickness not only in *Hakea* but other genera with large woody fruits [e.g *Corymbia* (Myrtaceae; Fig. 11.4); *Xylomelum* (Proteaceae)].

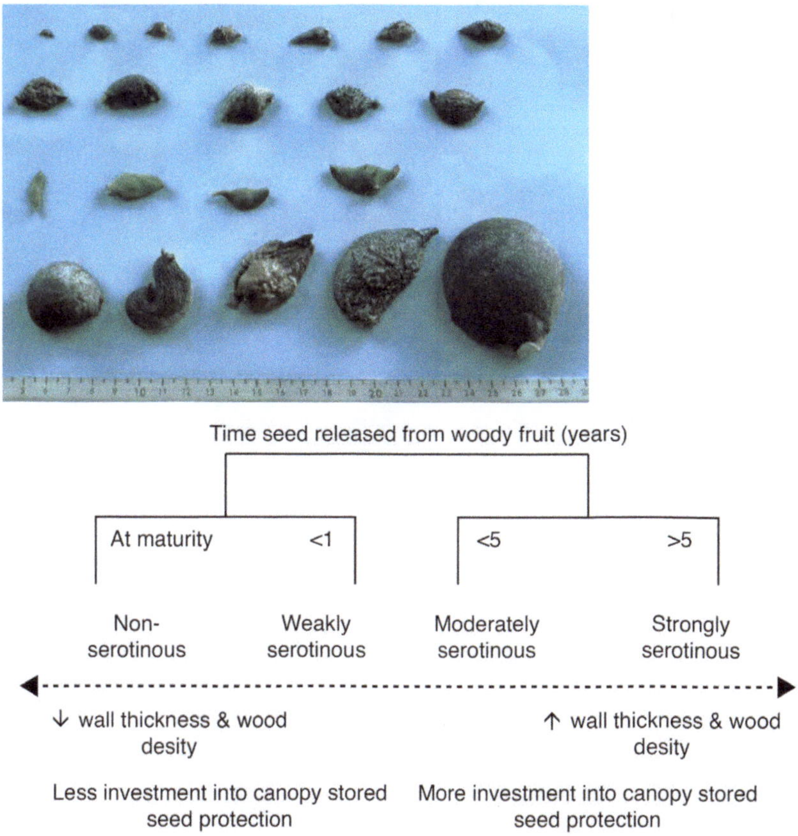

Fig. 11.3: (*upper*) Variability in fruit (follicle) size within SouthWest *Hakea* species, from the smallest (*H. sulcata*) to the largest (*H. platysperma*). Third row are species whose follicles remain green at maturity, and use cryptic mimicry as an additional form of protection (see Chapter 9). (*lower*) Fruit dimensions are related to level of seed protection in *Hakea*, with a trade-off between storage ability and fruit morphology, resulting in strongly serotinous species investing more into thicker-walled fruits than non- and weakly serotinous species. Diagram based on Groom & Lamont (1997).

Fig. 11.4: (*upper left*) Carnaby's black cockatoo (*Calyptorhynchus latirostris*), 35 cm tall, perched on a *Corymbia calophylla* (marri) branch. To the right of the branch is a single still-green woody fruit. When sectioned transversely (*upper right*) the thickness of the woody wall that protects the 3-4 seeds is apparent. Fruit diameter 5 cm. (*middle left*) The extraordinarily large fruits of *Hakea platysperma*, roughly two thirds the size of a Carnaby's cockatoo head (dummy head pictured – skull covered in plasticine – positioned against easily accessible fruits) provides sufficient woody protection to defend seeds against the bird's sharp bill. (*middle right*) Scratch marks on 6-cm-wide, *H. platysperma* follicle from failed attempts by a cockatoo to open the fruit. (*lower left*) In contrast, cockatoos easily extract seeds from smaller-fruited hakeas, even those with clustered fruits (e.g. *H. scoparia*). Fruit 2 cm long. (*lower right*) *Banksia attenuata* cones, 6 cm wide, removed from their parent stems by Carnaby's cockatoo. Note that subsequent damage is restricted to individual woody follicles. The seeds are even more vulnerable once the fruits have opened in response to fire, but remain protected by a woody separator that gradually lifts seeds out of the fruit.

The Calyptorhynchinae clade arose 22 Ma, early Miocene (White *et al.*, 2011), speciating from 15 Ma and becoming widespread in temperate Australia. *Calyptorhynchus* subgenus *Zanda* radiated 1.3 Ma resulting in two endemic species occupying today's SouthWest (Carnaby's and Baudin's cockatoos) and the Yellow-tailed in eastern Australia separated by the Nullarbor Plain that was by now a formidable barrier to migration. The third SouthWest *Calyptorhynchus* taxon is a subspecies of the Red-tailed black cockatoo that is more associated with tropical and arid Australia. At present only three black cockatoo species inhabit the SouthWest, with one critically endangered (Carnaby's cockatoo) and another endangered (forest red-tailed black cockatoo) because of extensive habitat destruction in both their breeding and feeding landscapes.

11.1.3 Post-fire Germination and Seedling Establishment

Seeds released from woody follicles and cones after fire are usually deposited into burnt litter microsites (Fig. 11.5) where they become buried by debris to a depth of 5–20 mm. For *Banksia* and *Hakea* species from the Eneabba sandplain, 340 km north of Perth, almost 100% germinate and there may be 120 to 450 seedlings m^{-2} by mid-winter (Lamont *et al.*, 1993). Over the ensuing summer-autumn, 67% of the seedlings die. That this reduction is largely due to competition for water in the highly packed microsites is indicated by only 27% dying if half the seedlings are removed before summer and the thinned seedlings are 59% larger than the undisturbed seedlings. Actually seeds that do not reach the litter microsites and remain in the sand patches are less likely to die (20%) because there are only 3 to 10 seedlings m^{-2} there. These patterns all help to explain the structure of the adult community after fire.

Fig. 11.5: Post-fire litter microsite full of seedlings produced from seeds that were blown there after their release from woody fruits following the fire. *Banksia* (10 cm tall) and needle-leaved *Hakea* seedlings dominate this microsite.

11.2 Soil-stored Seeds

Species that store seeds in the soil (geospory) account for 70-80% of the SouthWest flora, with the number and density of seeds stored differing between life form and habitat (Table 11.3). Seeds of fire ephemerals and long-lived perennials may persist in the soil for a few to many (e.g. 100) years, with seed dormancy broken in response to the occurrence of fire (heat and smoke cues), and/or changes in light, soil temperature or moisture conditions (Merritt *et al.*, 2007). Cues for breaking seed dormancy may also stimulate germination, but are often separate events in time as cues, especially fire, occur in summer while germination can only occur during the winter, when soil moisture is not limiting. Germination and seed dormancy requirements are reviewed in Merritt *et al.* (2007), and may be related to a species geographic range (Hidayati *et al.*, 2012). Fire-induced heat may be sufficient to break dormancy of soil-stored seeds with hard seed coats, especially legumes (Hanley & Lamont, 2000), provided they are not buried too deeply (usually < 5 cm).

Table 11.3: Soil-stored seed density in different SouthWest habitats. Life form categories are those listed in the respective publications.

	Seeds/m^2	Authors
Eneabba sandplains - scrub-heath		
Annuals	54-1,112	Enright *et al.* (2007)
Annuals	5	Bellairs & Bell (1993)
Perennials	149-310	Enright *et al.* (2007)
Perennials	43-110	Bellairs & Bell (1993)
Sedges	8-59	Bellairs & Bell (1993)
Exotics	3-71	Enright *et al.* (2007)
Exotics	3-5	Bellairs & Bell (1993)
Perth sandplains – Banksia **woodland**		
Native perennials	443	Roche *et al.* (1998)
Native annuals	347	Roche *et al.* (1998)
Exotics	18	Roche *et al.* (1998)
Jarrah forest		
Shrubs	1-22	Ward *et al.* (1997)
Sub-shrubs	1-142	Ward *et al.* (1997)
Herbs and annuals	1-19	Ward *et al.* (1997)
Wandoo woodland		
Grasses	30	Cochrane *et al.* (2007)
Sedges	6	Cochrane *et al.* (2007)
Herbs	196	Cochrane *et al.* (2007)
Exotics	21	Cochrane *et al.* (2007)

A combination of smoke followed by high soil summer temperatures has been shown to release seed dormancy in some soil-stored species (Tieu *et al.*, 2001a). Some SouthWest species known to germinate in response to smoke or heat treatments are listed in Table 11.4. The importance of smoke, usually applied as smoke water, as a germination stimulator for many species in fire-prone environments is now well established. The initial discovery was made by de Lange and Boucher (1990) on a shrub in the South African fynbos (shrublands similar to the SouthWest kwongan) and has subsequently been identified in other fire-prone environments including the Californian chaparral (Keeley & Fotheringham, 1998), Western Cape fynbos (Brown *et al.*, 2003) and the Mediterranean Basin (Crosti *et al.*, 2006).

The effectiveness of smoke in the SouthWest flora varies between families, genera, and species, due to the range of dormancy mechanisms that exists (Merritt *et al.*, 2007). There has been much research effort devoted to identifying the chemicals in smoke responsible for germination of soil-stored seeds (van Staden *et al.*, 2000; Light *et al.*, 2009). Smoke contains hundreds of compounds so it has not been easy. One of the two most active compounds discovered so far is glyceronitrile (H_3–CH(OH)–CN) that releases hydrogen cyanide (HCN) in the presence of water that stimulates germination (Flematti *et al.*, 2011; Downes *et al.*, 2013). The other is a butenolide-type compound, called karrikinolide (after *karri*, SouthWest aboriginal word for smoke) (Flematii *et al.*, 2004). Both are biologically active at extremely low concentrations (e.g. 1–10 μmol L^{-1}) (Table 11.5).

Anigozanthos (Haemodoraceae) is unresponsive to karrikinolide, but germinates in the presence of glyceronitrile (Flematti *et al.*, 2011; Downes *et al.*, 2013; Downes *et al.*, 2014). In contrast, the fire-ephemeral *Gyrostemon* (Gyrostemonaceae), which can only germinate in response to fire, germinates with smoke water but neither karrikinolide nor glyceronitrile is effective (Downes *et al.*, 2013). Another species in the same family, *Tersonia cyathiflora*, shows a similar response, but has not yet been tested with glyceronitrile (Downes *et al.*, 2010). *Blancoa canescens* (Haemodoraceae) seeds repond to karrikinoloide but not to glyceronitrile (Downes *et al.*, 2014). Such response variablility within the one family may provide insights into the evolutionary development of these smoke responses.

Table 11.4: Selection of SouthWest species whose germination has been shown to respond to smoke or heat. Fresh seeds were treated with cool smoke (derived from burning foliage of native plants) after storing in soil over summer-autumn and shown to germinate better than untreated fresh seeds (collated from Roche *et al.*, 1997). Species with hard seeds were immersed in water at 90°C and allowed to cool for 2 hours (collated from Cochrane *et al.*, 2002). Many species failed to germinate at all without smoke or heat treatment under otherwise ideal conditions.

Responsive to smoke			
Family	**Species**	**Family**	**Species**
Anthericaceae	*Agrostocrinum scabrum*	Myrtaceae	*Baeckea camphorosmae*
	Arthropodium strictum		*Calytrix fraseri*
	Johnsonia lupulina		*Calytrix tetragona*
	Sowerbaea multicaulis		*Verticordia aurea*
	Thysanotus multiflorus		*Verticordia chrysantha*
Apiaceae	*Actinotus helianthi*		*Verticordia densiflora*
	Xanthosia huegelii		*Verticordia eriocephala*
Apocynaceae	*Alyxia buxifolia*	Pittosporaceae	*Billardiera variifolia*
Asteraceae	*Myriocephalus stuartii*	Poaceae	*Hovea trisperma*
Cyperaceae	*Gahnia lanigera*	Phormiaceae	*Dianella revoluta*
Dasypogonaceae	*Acanthocarpus preissii*	Polygalaceae	*Comesperma virgatum*
	Lomandra preissii	Proteaceae	*Conospermum incurvum*
Dilleniaceae	*Hibbertia amplexicaulis*		*Conospermum triplinervium*
	Hibbertia commutata		*Grevillea polybotrya*
	Hibbertia lasiopus		*Grevillea quercifolia*
	Hibbertia ovata		*Grevillea scapigera*
	Hibbertia quadricolor		*Grevillea wilsonii*
	Hibbertia riparia		*Stirlingia latifolia*
	Hibbertia sericea	Ranunculaceae	*Clematis pubescens*
	Hibbertia serrata	Rutaceae	*Boronia fastigiata*
Ecdeiocoleaceae	*Georgeantha hexandra*		*Boronia megastigma*
Ericaceae	*Acrotriche patula*		*Boronia viminea*
	Andersonia involucrata		*Diplolaena dampieri*
	Astroloma foliosum		*Geleznowia verrucosa*
	Astroloma pallidum	Santalaceae	*Exocarpus sparteus*
	Leucopogon capitellatus		*Leptomeria cunninghamii*
	Leucopogon nutans	Stackhousiaceae	*Stackhousia pubescens*
	Leucopogon propinquus		*Tripterococcus brunonis*
	Leucopogon verticillatus	Stylidiaceae	*Stylidium affine*
Euphorbiaceae	*Ricinocarpos glaucus*		*Stylidium amoenum*
Goodeniaceae	*Scaevola calliptera*		*Stylidium bulbiferum*

continued**Table 11.4:** Selection of SouthWest species whose germination has been shown to respond to smoke or heat. Fresh seeds were treated with cool smoke (derived from burning foliage of native plants) after storing in soil over summer-autumn and shown to germinate better than untreated fresh seeds (collated from Roche *et al.*, 1997). Species with hard seeds were immersed in water at 90°C and allowed to cool for 2 hours (collated from Cochrane *et al.*, 2002). Many species failed to germinate at all without smoke or heat treatment under otherwise ideal conditions.

Responsive to smoke

Family	Species	Family	Species
	Scaevola fasciculata		*Stylidium hispidum*
Gyrostemonaceae	*Tersonia cyathiflora*		*Stylidium junceum*
Haemodoraceae	*Anigozanthos manglesii*		*Stylidium schoenoides*
	Anigozanthos rufus	Thymelaeaceae	*Pimelea ciliata*
	Blancoa canescens		*Pimelea imbricata*
	Conostylis aculeata		*Pimelea leucantha*
	Macropidia fuliginosa		*Pimelea suaveolens*
Iridaceae	*Orthrosanthus laxus*	Tremandraceae	*Tetratheca hirsuta*
	Patersonia glabrata	Violaceae	*Hybanthus calycinus*
	Patersonia occidentalis		
	Patersonia aff. *fragilis*		

Responsive to heat

Family	Species	Family	Species
Fabaceae;	*Chorizema humile*	Fabaceae;	*Acacia awestoniana*
Faboideae	*Daviesia campephylla*	Mimosoideae	*Acacia brachypoda*
	Daviesia elongata ssp. *elongata*		*Acacia cochlocarpa* ssp. *velutinosa*
	Daviesia euphorbioides		*Acacia heteroclita* ssp. *valida*
	Daviesia megacalyx		*Acacia leptalea*
	Daviesia microcarpa		*Acacia pygmaea*
	Daviesia pseudaphylla		*Acacia sciophanes*
	Daviesia spiralis		*Acacia sp. Dandaragan*
	Gastrolobium papilio		*Acacia subflexuosa* ssp. *capitellata*
	Jacksonia pungens	Malvaceae;	*Rulingia* sp. Trigwell Bridge
	Jacksonia quairading	Sterculioideae	*Thomasia* sp. Green Hill
	Kennedia glabrata		
	Kennedia macrophylla		
	Nemcia aff. *rubra*		

Table 11.5: Germination response (%) of SouthWest species with two of the known active compounds present in smoke water. Note that the responses vary greatly between species, with some only responsive to glyceronitrile (*A. manglesii*) or karrikinolide (*S. affine*) or both (*C. candicans*), or neither (*A. flavidus*). *Tersonia cyathiflora* is particularly interesting as it has an obligate requirement for smoke (so that the species must have evolved with fire) but the active compound has not yet been discovered, though it might be the untried glyceronitrile. Summarized from Flematti *et al.* (2011) and Downes *et al.* (2010).

Species (family)	Control (water only)	Smoke water (diluted 1/10)	Karrikinolide (1 μM)	Glyceronitrile (50 μM)
Anigozanthos manglesii (Haemodoraceae)	21	44	16	52
Stylidium affine (Stylidiaceae)	7	80	84	31
Conostylis candicans (Haemodoraceae)	51	70	75	79
Anigozanthos flavidus (Haemodoraceae)	66	74	67	84
Tersonia cyathiflora (Gyrostemonaceae)	0	67	0	–

11.2.1 Advantages of Storing Seeds in the Soil

Soil-stored seeds are rarely present at depths greater than 10 cm from the soil surface, with the greatest seed density occurring in the top 5 cm. Storing seeds in the soil is advantageous because it is a microclimate that provides some protection from granivores and insulates the seed from the direct impact of a surface fire (Tieu *et al.*, 2000; Tieu *et al.*, 2001b). Seed burial is a requirement for breaking dormancy in many SouthWest species, and may involve other cues, such as the occurrence of wet-dry cycles (promoting seeds to germinate at the beginning of the wet season), strong daily temperature fluctuations (as occurs after removal of the plant cover by fire) or exposure to smoke (promoting germination post-fire). Burial may cause a gradual decay of the seed coat, allowing the seed to imbibe water during the next rain event (Baker *et al.*, 2005a; 2005b). Fire-induced heat is often required to crack the seed coat, or dislodge the strophiolar plug (Dell, 1980), of many hard-seeded soil-stored species to achieve the same effect.

The main advantage over serotiny is that usually all stored seeds do not germinate in response to a given fire as they may escape the heat pulse so may germinate at a later time (Fig. 11.6). This may be adaptive if recruitment conditions are poor so that all the seeds are not wasted (called 'bet-hedging'). For serotinous species, it is essential that seed release be followed by a reliable wet season as all seeds are released in non-patchy fires. As it turns out, the SouthWest has the most reliable winter among all five mediterranean regions of the world (Cowling *et al.*, 2005) and can explain why serotiny is also best represented here.

Fig. 11.6: Young plants of *Paraserianthes lophantha* (Fabaceae, Mimosoideae) following burning of a stockpile of plant debris in the Darling Range near Perth. Seeds of this species already lay buried in the surface soil and were stimulated to germinate by the heat of the fire. Note how the seedlings form a ring around the stockpile: heat was excessive and killed seeds in the centre, the heat pulse was insufficient to stimulate germination beyond the ring and the seeds remained dormant, though they could germinate at a later date with a subsequent fire ('bet-hedging'). In a similar way, depth of burial is critical to whether seeds germinate, are killed (too shallow) or remain dormant (too deep). Typical of legumes, this species is taking full advantage of the mineral nutrients released by the fire. Photo taken by B. Dell.

11.2.2 Case Study: *Austrostipa compressa* and *Ehrharta calycina*

Austrostipa compressa is a fire ephemeral, native annual grass that stores most of its seeds near the soil surface (< 5 cm depth) at a density ranging from 45 to 6,000 seeds m^{-2}, depending on time since last fire (Smith *et al.*, 1999). The surface soil contains at least 80% of the germinable seedbank, with 70% seed viability. In recently burnt sites, the greatest number of seeds germinates during the first winter rains in response to smoke cues. Seeds of *A. compressa* are capable of burying themselves 2–4 cm below the soil surface through the corkscrew action of their awns, and it is these darkened conditions that maintain the seeds in a dormant state in the soil between fires. Exposure to light is also a germination stimulus, providing the seeds with the opportunity to germinate in response to soil disturbance.

The exotic perennial grass, *Ehrharta calycina* (perennial veldtgrass), has been recorded with a soil-stored seed density up to 52,000 seeds m^{-2} in the litter with up to 21,000 seeds m^{-2} in the surface soil (< 5 cm). *E. calycina* seeds are not as deeply buried as those of *A. compressa*, mostly lying on the soil surface, and hence will be consumed by a passing fire. Germination of *E. calycina* seeds is not enhanced in response to smoke or smoke water, with germination cued only by the availability of adequate moisture. *E. calycina* produces large number of seeds annually with a high probability of viable seeds germinating in the following year, explaining why it has become a major weed on the Swan Coastal Plain.

11.3 Seed Size and Seedling Survival

The best opportunity for seedling survival is the first wet season after fire when competition from parents is minimal, the occurrence of another fire before reaching maturity is unlikely, and light, nutrients and water are at a temporary maximum. Thus, serotinous seeds are cued for release from their fruiting bodies by fire heat directly or following death of the supporting stems and subsequent wet-dry cycles, and soil-stored seeds are stimulated to germinate by fire heat or smoke, or subsequent increased daily soil temperature fluctuations. Nevertheless, most seedlings are crowded into post-fire litter microsites around the parent plant or ant 'middens' where interplant competition for nutrients and water is intense (Lamont *et al.*, 1993). In addition, much of the litter nutrients is lost in smoke, decomposing litter as a source of mineral nutrients is temporally halted, and nutrient levels remain scarce by world standards; while a severe summer–autumn drought, which completely dries out the surface soil within 6 months of germination, is imminent.

There are three possible adaptations to ensure survival of the population after fire: 1) adults resprout (Chapter 1, 2), adults die but regenerate from many small seeds, or 3) adults die but regenerate from a few large seeds (Richards & Lamont, 1996; Esther *et al.*, 2011). But even resprouters must produce successful seedlings at some stage (unless they are clonal) or the population will gradually decline as immature plants not yet fire-tolerant are killed by fire and old plants eventually die. And only seeds give the opportunity for a genetic response to changing growing conditions over geological time (though clonal species that experience beneficial somatic mutations could grow preferentially in that part of the clone (Wiens & Lamont, 2003)). Extensive demographic studies on hakeas and banksias in the SouthWest have shown that post-fire recruitment patterns conform to biased lottery models, with the *demographic* (lottery) component, fecundity (seeds per parent), the key for survival, followed closely by the *biotic* (biased) component, especially seed size (Lamont *et al.*, 1999).

Demography refers to the number of individuals present over time, while a lottery is where the chance of success depends on how many individuals of a given type are represented in a population: thus, the more seeds stored per parent, the more seedlings produced post-fire and the more seedlings are recruited into the new community.

Fig. 11.7: 7-week-old seedlings of three *Hakea* species endemic to the SouthWest at the same scale (25-mm rim of pot visible in background) showing the range of variation in leaf shape and cotyledon size in this genus. (*left*) *H. sulcata* has the smallest seeds of all 140 hakeas (2.7 mg) and thus the smallest cotyledons (the first pair of 'leaves' just visible) and seedlings—its broad leaves are replaced by simple needle leaves once the eighth leaf is produced (heteroblasty). This species is confined to seasonal wetlands and this may enhance its ability to establish a strong root system before the summer drought commences by which time it only produces vertically-oriented needle leaves. (*middle*) *H. lissocarpha* has intermediate-sized seeds (23.9 mg) and occurs on shallow lateritic soils. The nutrient store is sufficient to ensure development of a deep root system by summer plus its semi-terete, highly-divided, needle leaves increase its drought tolerance. (*right*) *H. platysperma* has the largest seeds of all hakeas (508.8 mg) and thus the largest seedlings. Its cotyledons expand four times in area on germination and become strongly photosynthetic. The seeds contain more P (15 mg) than is known for any other species (Groom & Lamont, 2010). Thus, the major role of its cotyledons is the provision of mineral nutrients to the shoots that enhance photosynthesis and enable rapid elongation of the tap root in the deep sands where it occurs to ensure contact continues with groundwater over summer-autumn (Fig. 10.15, Lamont & Groom, 2013). Drought avoidance is assisted from the outset by its exceptionally-long, vertically-oriented, needle leaves.

11.3.1 Small Seeds Produce Small Seedlings

Small seeds produce small seedlings, so how do they survive? By producing many seeds, more will be dispersed further from the plant to be deposited in 'safe' sites, such as bare patches free of competitors (Lamont *et al.*, 1993), by chance alone (demographic explanation). Some small seeds produce seedlings that are more drought tolerant, e.g. they survive at relative water contents 35–45% below the turgor loss point (Richards & Lamont, 1996) or are needle- rather than broad-leaved (Lamont & Groom, 1998) so minimizing rates of transpiration (biotic explanation). Nevertheless, overall, seedlings are more likely to survive the first summer drought, the larger the seeds from which they arose, although at a rapidly diminishing rate (Lamont & Groom, 2013). The mechanism starts with nutrient accumulation in the seeds. Nonsprouters in particular allocate much of their aboveground nutrient store to seeds. For example, the seeds of *Banksia hookeriana* in the Eneabba sandplains (Fig. 11.2) account for 0.5% of shoot weight but contain 48% of its P (three times that in the leaves), 24% of its N (the same as in the leaves) and 4% of its potassium (Witkowski & Lamont, 1996). Usually the endosperm is absorbed (rarely it remains, as in cycads) and nutrients are stored in the cotyledons. Generally, the cotyledons are carried above the soil during germination (epigeal), they spread and become photosynthetic (Fig. 11.7).

Removing the cotyledons from seedlings produced by small seeds has little effect on their growth while adding nutrients increases their growth by many times (Milberg & Lamont, 1997; Lamont & Groom, 1998). By contrast, the growth of cotyledon-removed seedlings from seeds > 20 mg weight is retarded markedly, they may even die, while in their presence they grow just as well whether they are given extra nutrients or not (Milberg *et al.*, 1998). Adding a balanced nutrient solution to the minus-cotyledon seedlings returns their growth to that of the plus-cotyledon seedlings (Milberg & Lamont, 1997; Lamont & Groom, 2002). It is clear that many species in the SouthWest, especially the serotinous ones, have opted for large seeds (Groom & Lamont, 2010) and that such sizes are exceptional at the regional level, especially when compared with other temperate southern hemisphere floras (Fig. 11.8).

The greater early growth from large seeds (Fig. 11.9) can be attributed to the efficient transport of nutrients from the cotyledons to the seedling. For *Hakea psilorrhyncha*, with a 67-mg seed, 97% of its P is transferred, 85% of its N, 90% of its K, 91% of its copper, 71% of its iron, 50% of its manganese and 29% of its calcium (Lamont & Groom, 2002). As a result, 79% of P in the seedling by 12 weeks is due to import from the cotyledons down to 3% of calcium. Calcium, magnesium and iron are more likely to be supplied from ash after fire. These retranslocated nutrients are mostly shunted to the young shoot where they are vital for photosynthesis, respiration and transpiration. The fixed carbon (C) is used to produce new roots (in particular) and leaves and stem elongation. It is rapid and deep extension of the tap root, rather than extensive lateral root growth that is the key to survival over the long summer-autumn drought. The tap root from large-seeded species can be traced to

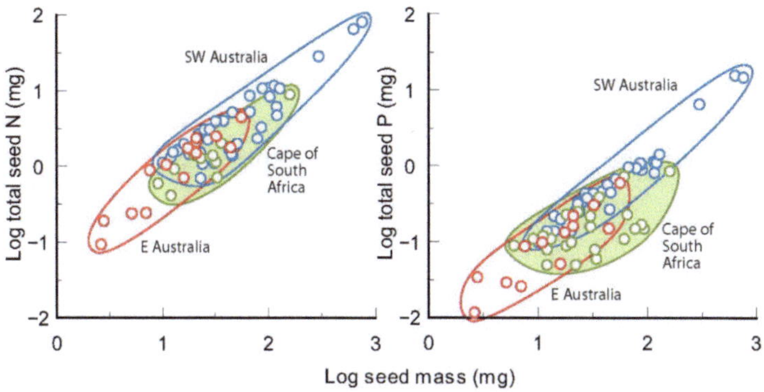

Fig. 11.8: The Proteaceae occurs widely in Australia and South Africa and provides the opportunity to compare seed sizes and their nutrient contents between regions. As can be seen, seeds are generally larger (by an average of 9 times) and more enriched with N (by an average of 13 times) and P (by an average of 30 times) in the SouthWest than on the east side of Australia, with mediterranean South Africa in between. From Lamont & Groom (2013).

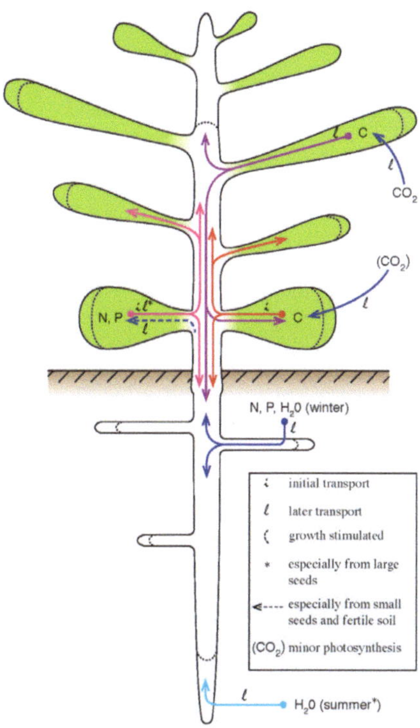

Fig. 11.9: Stylized epigeal seedling showing uptake and transport of carbon (C), nitrogen (N), phosphorus (P) and water from soils in the SouthWest. Note that the cotyledons are short-lived as a source of C as they soon become a sink for C following photosynthesis by the leaves. Except for small seeds and fertile soils, N and P continue to be transported out of the cotyledons until their supply is exhausted. This ready source of N and P in nutrient-limited soils enhances photosynthesis and general growth. Fixed C is shunted strongly to the root system where N, P and water uptake and elongation of the tap root in particular are promoted, as an indirect effect of greater N and P seed stores. From Lamont & Groom (2013), figure prepared by Wesley Lamont.

a depth of 2 m or more, up to ten times the stature of the plant, within 10 months of germination (Enright & Lamont, 1992; Rokich *et al.*, 2001). The close relationship between abundant light, high nutrient storage/transport and subsequent drought resistance (avoidance) hold the key to understanding the fitness advantage of large seeds in the SouthWest.

Bibliography

Ackerly, D. D., & Cornwell, W. K. (2007). A trait-based approach to community assembly: portioning of species trait values into within- and among-community components. *Ecology Letters, 10,* 135-145.

Ackerman, J D. (1986). Mechanisms and evolution of food-deceptive pollination systems in orchids. *Lindleyana, 1,* 108-113.

Alcock, J. (2000). Interactions between the sexually deceptive orchid *Spiculaea ciliata* and its wasp pollinator *Thynnoturneria* sp. (Hymenoptera: Thynninae). *Journal of Natural History, 34,* 629-636.

Almeida, E. A. B., Pie, M. R., Brady, S. G, and Danforth, B. N. (2012). Biogeography and diversification of colletid bees (Hymenoptera: Colletidae): emerging patterns from the southern end of the world. *Journal of Biogeography, 39,* 526-544.

Armbruster, W. C., Edwards, M. E., & Debevec, E. M. (1994). Floral character displacement generates assemblage structure of Western Australian triggerplants (*Stylidium*). *Ecology,* 75, 315-329.

Armstrong, J. A., (1979). Biotic pollination mechanisms in the Australian flora—a review. *New Zealand Journal of Botany, 17,* 467-508.

Auld, T. D. (2009). Petals may act as a reward: myrmecochory in shrubby *Darwinia* species of south-eastern Australia. *Austral Ecology, 34,* 351-356.

Australia's Bioregions. http://www.environment.gov.au/topics/land/national-reserve-system/science-maps-and-data/australias-bioregions-ibra. Accessed 1 October 2014

Bader, M. K. F., Ehrenberger, W., Bitter, R., Stevens, J., Miller, B.P., Chopard, J., Rüger, S., Hardy, G. E. S. J., Poot, P., Dixon, K. W., Zimmermann, U., & Veneklaas, E. J. (2014). Spatio-temporal water dynamics in mature *Banksia menziesii* trees during drought. *Physiologia Plantarum, 152,* 301-315.

Baker, A. J., Haddrath, McPherson, & Cloutier, A. (2014). Genomic support for a moa-tinamou clade and adaptive morphological convergence in flightless ratites. *Molecular Biology and Evolution,* 31, 1686-1696.

Baker, K. S., Steadman, K. J., Plummer, J. A., Merritt, D. J., & Dixon, K. W. (2005a). The changing window of conditions that promotes germination of two fire ephemerals, *Actinotus leucocephalus* (Apiaceae) and *Tersonia cyathiflora* (Gyrostemonaceae). *Annals of Botany, 96,* 1225-1236.

Baker, K. S., Steadman, K. J., Plummer, J. A., Merritt, D. J., & Dixon, K. W. (2005b). Dormancy release in Australian fire ephemeral seeds during burial increases germination responses to smoke water or heat. *Seed Science Research, 15,* 339-348.

Barkman, T. J, McNeal, J. R, Lim, S.-H., Coat, G., Croom, H. B., Young, N. D., & de Pamphilis, C. W. (2007). Mitochondrial DNA suggests at least 11 origins of parasitism in angiosperms and reveals genomic chimerism in parasitic plants. *BMC Evolutionary Biology, 7,* 248

Barrett, R. L. (2013). Ecological important of sedges: a survey of the Australasian Cyperaceae genus *Lepidosperma. Annals of Botany, 111,* 499-529.

Barron, O., Froend, R., Hodgson, G., Ali, R., Dawes, W., Davies, P., & McFarlane, D. (2014). Projected risks to groundwater-dependent terrestrial vegetation caused by changing climate and groundwater abstraction in the Central Perth Basin, Western Australia. *Hydrological Processes, 28,* 5513-5529.

Batley, M., & Hogendoorn, K. (2009). Diversity and conservation status of native Australian bees. *Apidologie, 40,* 347-354.

Batty, A. L., Dixon, K. W., Brundrett, M., & Sivasithamparam, K. (2001). Contraints to symbiotic germination of terrestrial orchid seed in a mediterranean bushland. *New Phytologist, 152,* 511-520.

Batty, A. L., Dixon, K. W., & Sivasithamparam, K. (2000). Soil-seed bank dynamics of terrestrial orchids. *Lindleyana, 15,* 227-236.

Beard, J. S. (1980). A new phytogeographic map of Western Australia. *Western Australian Herbarium Research Notes, 3,* 37-58.

Beard, J. S. (1984). Biogeography of the kwongan. In J. S. Pate and J. S. Beard (Eds), *Kwongan. Plant life of the sandplain.* pp. 1-26. Nedlands, Australia: University of Western Australia -Press.

Beard, J. S. (1990). *Plant life of Western Australia.* Kenthurst, Australia: Kangaroo Press.

Beard, J. S., Chapman, A. R., & Gioia, P. (2000). Species richness and endemism in the Western Australian flora. *Journal of Biogeography, 27,* 1257-1268.

Beardsell, D. V., Clements, M. A., Hutchinson, J. F., & Williams, E. G. (1986). Pollination of *Diuris maculata* R Br (Orchidaceae) by floral mimicry of the native legume *Daviesia* spp and *Pultenaea scabra* R Br. *Australian Journal of Botany, 34,* 165-173.

Beardsell, D. V., O'Brien, S. P., Williams, E. G., Knox, R. B, & Calder, D. M. (1993). Reproductive biology of Australian Myrtaceae. *Australian Journal of Botany, 41,* 511-526.

Bell, D. T. (2001). Ecological response syndromes in the flora of southwestern Western Australia: fire resprouters versus reseeders. *The Botanical Review, 67,* 417-440.

Bellairs, S. M., & Bell, D. T. (1993). Seed stores for restoration of species-rich shrubland vegetation following mining in Western Australia. *Restoration Ecology, 1,* 233-241.

Bellamy, C. L., Williams, G. A., Hasenpusch, J., & Sundholm, A. (2013). A summary of the published data on host plants and morphology of immature stages of Australian jewel beetles (Coleoptera: Buprestidae), with additional new records. *Insecta Mundi, 293,* 1-172.

Berg, R. Y. (1975). Myrmecochorous plants in Australia and their dispersal by ants. *Australian Journal of Botany, 23,* 475-508.

Bernhardt, P. (1987). A comparison of the diversity, density, and foraging behaviour of bees and wasps on Australian *Acacia*. *Annals of the Missouri Botanic Garden, 74,* 42-50.

Bernhardt, P. (1995). Notes on the anthecology of *Pterostylis curta* (Orchidaceae). *Cunninghamia, 4,* 1-8.

Bernhardt, P., & Burns-Balogh, P. (1986a). Observations of the floral biology of *Prasophyllum odoratum* (Orchidaceae, Spiranthoideae). *Plant Systematics and Evolution, 153,* 65-76.

Bernhardt, P., & Burns-Balogh, P. (1986b). Floral mimesis of *Thelymitra nuda* (Orchidaceae). *Plant Systematics and Evolution, 151,* 187-202.

Blanck, Y. L., Gowda, J., Martensoon, L. M., Sandberg, J., & Fransson, A. M. (2011). Plant species richness in a natural Argentinean matorral shrub-land correlates negatively with levels of plant phosphorus. *Plant and Soil, 345,* 11-21.

Bohman, B., & Peakall, R. (2014). Pyrazines attract *Catocheilus* thynnine wasps. *Insects, 5,* 474-487.

Bohman, B., Phillips, R. D., Menz, M. H. M., Bernstsson, B. W., Flematti, G. R., Barrow, R. A., Dixon, K. W., & Peakall, R. (2014). Discovery of pyrazines as pollinator sex pheromones and orchid semiochemicals: implications for the evolution of sexual deception. *New Phytologist, 203,* 939-952.

Bond, W. J., & Keeley, J. E. (2005). Fire as a global 'herbivore': the ecology and evolution of flammable ecosystems. *Trends in Ecology and Evolution, 20,* 387-394.

Bond, W. J., & Scott, A. C. (2010). Fire and the spread of flowering plants in the Cretaceous. *New Phytologist, 188,* 1137-1150.

Bougoure, J., Brundrett, M., Brown, A., & Grierson, P. F. (2008). Habitat characteristics of the rare underground orchid *Rhizanthella gardneri*. *Australian Journal of Botany, 56,* 501-511.

Bougoure, J. J., Brundrett, M. C., & Grierson, P. F. (2010). Carbon and nitrogen supply to the underground orchid, *Rhizanthella gardneri*. *New Phytologist, 186,* 947-956.

Bougoure, J., Ludwig, MK., Brundrett, M., & Grierson, P. (2009). Identity and specificity of the fungi forming mycorrhizas with the rare mycoheterotrophic orchid *Rhizanthella gardneri*. *Mycological Research, 113,* 1097-1106.

Bower, C. C. (2014). Pollination of the small duck orchid, *Paracaleana minor*: flower structure and function. *The Orchardian, 17,* 510-515.

Bradshaw, S. D., & Bradshaw, F. J. (1999). Field energetics and the estimation of pollen and nectar intake in the marsupial honey possum, *Tarsipes rostratus,* in heathland habitats of south-western Australia. *Journal of Comparative Physiology B, 169,* 569-580.

Bradshaw, S. D., & Bradshaw, F. J. (2002). Short term movements and habitat use of the marsupial honey possum (*Tarsipes rostratus*). *Journal of Zoology, 258,* 343-348.

Bradshaw, S. D., Phillips, R. D., Tomlinson, S., Holley, R. J., Jennings, S., & Bradshaw, F. J. (2007). Ecology of the honey possum, *Tarsipes rostratus,* in Scott National Park, Western Australia. *Australian Mammalogy, 29,* 25-38.

Bradstock, R. A., Gill, A. M., Hastings, S. M., & Moore, P. H. R. (1994). Survival of serotinous seedbanks during bushfires: comparative studies of *Hakea* species from southeastern Australia. *Australian Journal of Ecology, 19,* 276-282.

Bremer, K. (2002). Gondwanan evolution of the grass alliance of families (Poales). Evolution 56, 1374–1387.

Briggs, B. G., Marchant, A., Gilmore, S., & Porter, C. (2000). A molecular phylogeny of Restionaceae and allies. In K. L. Wilson & D. A. Morrison (Eds), *Systematics and evolution of monocots.* Vol. 1. Proceedings of the 2nd International Conference on the Comparative Biology of the Monocots, Sydney. pp. 661–671. Melbourne: CSIRO Publishing.

Brown, A. P., Dundas, P., Dixon, K., & Hopper, S. D. (2008). *Orchids of Western Australia.* Nedlands, Australia, University of Western Australia Press.

Brown, N. A. C., van Staden, J., Daws, M. I., & Johnson, T. (1993). Patterns in the seed germination response to smoke in plants from the Cape Floristic Region, South Africa. *South African Journal of Botany, 69,* 514-525.

Brundrett, M. (2004). Diversity and classification of mycorrhizal associations. *Biological Reviews, 79,* 473–495.

Brundrett, M. C. (2007). Scientific approaches to Australian temperate terrestrial orchid conservation. *Australian Journal of Botany, 55,* 293-307.

Brundrett, M. C. (2009). Mycorrhizal associations and other means of nutrition of vascular plants: understanding the global diversity of host plants by resolving conflicting information and developing reliable means of diagnosis. *Plant and Soil, 320,* 37-77.

Brundrett, M. C., & Abbott, L. K. (1991). Roots of jarrah forest plants. I. Mycorrhizal associations of shrubs and herbaceous plants. *Australian Journal of Botany, 39,* 445-457.

Burbidge, A. H., & Whelan, R. J. (1982). Seed dispersal in a cycad, *Macrozamia riedlei. Australian Journal of Ecology, 7,* 63-67.

Burgess, S. S. O., Adams, M. A., Turner, N. C., White, D. A., & Ong, C. K. (2001). Tree roots: conduits for deep recharge of soil water. *Oecologia, 126,* 158-165.

Burgess, S. S. O., Pate, J. S., Adams, M. A., & Dawson, T. E. (2000). Seasonal water acquisition and redistribution in the Australian woody phreatophyte, *Banksia prionotes. Annals of Botany, 85,* 215-224.

Burrows, G. E. (2013). Buds, bushfires and resprouting in the eucalypts. *Australian Journal of Botany, 61,* 331-349.

Burrows, N. D., & McCaw, W. L., (1990). Fuel characteristics and bushfire control in banksia low woodlands in Western Australia. *Journal of Environmental Management, 31,* 229-236.

Byrne, M. (2008). Evidence for multiple refugia at different time scales during Pleistocene climatic oscillations in southern Australia inferred from phylogeography. *Quaternary Science Reviews, 27,* 2576-2585.

Byrne, M., & Hopper, S. D. (2008). Granite outcrops as ancient islands in old landscapes: evidence from the phylogeography and population genetics of *Eucalyptus caesia* (Myrtaceae) in Western Australia. *Biological Journal of the Linnean Society, 93,* 177–188

Byrne, M., Elliott, C. P., Yates, C., & Coates, D. J. (2007). Extensive pollen dispersal in a bird-pollinated shrub, *Calothamnus quadrifidus*, in a fragmented landscape. *Molecular Ecology*, 16, 1303-1314.

Byrne, M., Yeates, D.K., Joseph, L., Kearney, M., Bowler, J., Williams, M. A.J., Cooper, S., Donnellan, S.C., Keogh, J.S., Leys, R., Melville, J., Murphy, D.J., Porch, N., & Wyrwoll, K.-H. (2008). Birth of a biome: insights into the assembly and maintenance of the Australian arid zone biota. *Molecular Ecology, 17,* 4398–4417.

Byrne, M., Steane, D. A., Joseph, L., Yeates, D. K, Jordan, G. J., Crayn, D., Aplin, K., Cantrill, D. J., Cook, L. G., Crisp, M. D., Keogh, J. S., Melville, J., Moritz, C., Porch, N., Sniderman, J. M. K., Sunnucks, P., & Weston, P. H. (2011). Decline of a biome: evolution, contraction, fragemention, extinction and invasion of the Australian mesic zone biota. *Journal of Biogeography, 38,* 1635-1656.

Byrne, M., Coates, D. J., Forest, F., Hopper, S. D., Krauss, S. L., Kale Sniderman, J. M.,, & Thiele, K. R. (2014). A diverse flora – species and genetic relationships In H. Lambers (Ed), *Plant life on the sandplains in southwest Australia.* pp. 81-100. Perth, Australia: UWA Publishing.

Calladine, A., & Pate, J. S. (2000). Haustorial structure and functioning of the root hemiparastic tree *Nuytsia floribunda* (Labill.) R.Br. and water relationships with its hosts. *Annals of Botany, 85,* 723-731.

Calviño-Cancela, M., Dunn, R. R., van Etten, E. J. B., & Lamont, B. B. (2006). Emus as non-standard seed dispersers and their potential for long-distance dispersal. *Ecography, 29,* 632-640.

Calviño-Cancela, M., He, T., & Lamont, B. B. (2008). Distribution of myrmecochorous species over the landscape and their potential long-distance dispersal by emus and kangaroos. *Diversity and Distributions, 14,* 11-17.

Canadell, J., Jackson, R. B., Ehleringer, J. R., Mooney, H. A., Sala, O. E., & Schulze, E. -D. (1996). Maximum rooting depth of vegetation types at the global scale. *Oecologia, 108,* 583-595.

Canham, C. A., Froend, R. H., & Stock, W. D. (2009). Water stress vulnerability of four *Banksia* species in contrasting ecophydrological habitats on the Gnangara Mound, Western Australia. *Plant, Cell and Environment, 32,* 64-72.

Canham, C. A., Froend, R. H., Stock, W. D., & Davies, M. (2012). Dynamics of phreatophyte root growth relative to a seasonally fluctuating water table in a Mediterranean-type environment. *Oecologia, 170,* 909-915.

Case, A. L., & Barrett, S. C. H. (2004).Environmental stress and the evolution of dioecy: *Wurmbea dioica* (Colchicaceae) in Western Australia. *Evolutionary Ecology*, 81, 145-164.

Chalwell, S. T. S., & Ladd, P. G. (2005). Stem demography and post fire recruitment of *Podocarpus drouynianus*: a resprouting non-serotinous conifer. *Botanical Journal of the Linnaean Society, 149,* 433–449.

Chandler, G. T. Crisp, M. D., Cavzer, L. W., & Bayer, R. J. (2002). Monograph of *Gastrolobium* (Fabaceae: Mirbelieae). *Australian Systematic Botany, 15,* 619-739.

Chinook, R. J. (2007). *Eremophila and allied genera: a monograph of the plant family Myoporaceae.* Kenrhurst, NSW, Australia: Rosenberg Publishing.

Chittka, L., Thomson, J. D., & Waser, N. M. (1999). Flower constancy, insect psychology, and plant evolution. *Naturwissenschaften, 86,* 361–377.

Choudhury, N.K., & Sahu, D. (1999). Photosynthesis in *Cuscuta reflexa*: a total plant parasite. *Photosynthetica, 36,* 1-9.

Claridge, A. W. (2002). Ecological role of hypogeous ectomycorrhizal fungi in Australian forests and woodlands. *Plant and Soil, 244,* 291-305.

Clarke, J. T., Warnock, R. C. M., & Donoghue, P.C.J. (2011). Establishing a time-scale for plant evolution. *New Phytologist, 192,* 266-301.

Clarke, P. J., Lawes, M. J., Midgley, J. J., Lamont, B. B., Ojeda, F., Burrows, G. E., Enright, N. J., & Knox, K. J. E. (2013). Resprouting as a key functional trait: how buds, protection and resources drive persistence after fire. *New Phytologist, 197,* 19-35.

Cochrane, A., Kelly, A., Brown, K., & Cunneen, S. (2002). Relationships between seed germination requirements and ecophysiological characteristics aid the recovery of threatened native plant species in Western Australia. *Ecological Management and Restoration, 3,* 47-60.

Cochrane, A., Monks, L., & Lally, T. (2007). Response of the germinable soil-stored seed bank of a remnant reserve in the southern Western Australian agricultural zone to smoke and fire treatment. *Journal of the Royal Society of Western Australia*, 90, 47-52.

Collins, B. G. (1981). Nectar intake and water balance for two species of Australian honeyeater, *Lichmera indistincta* and *Acanthorhynchus superciliosis*. *Physiological Zoology, 54,* 1-13.

Collins, B. G. (1985). Energetics of foraging and resource selection by honeyeaters in forest and woodland habitats of Western Australia. *New Zealand Journal of Ecology, 12,* 577-587.

Collins, B. G., & Briffa, P. (1983). Seasonal and diurnal variations in the energetics and foraging activities of the brown honeyeater, *Lichmera indistincta*. *Australian Journal of Ecology,* 8, 103-111.

Collins, B. G., & Morellini, P. C. (1979). The influence of nectar concentration and time of day upon energy intake and expenditure by the singing honeyeater, *Meliphaga virescens*. *Physiological Zoology, 52,* 165-175.

Collins, B. G., Newland, C., & Briffa, P. (1984). Nectar utilization and pollination by Australian honeyeaters and insects visiting *Calothamnus quadrifidus* (Myrtaceae). *Australian Journal of Ecology, 9,* 353-365.

Collins, B. G. & Spice, J. (1986). Honeyeaters and the pollination biology of *Banksia prionotes* (Proteaceae). *Australian Journal of Botany, 34,* 175-185.

Conran, J. G. (2008). Aestivation organ structure in *Drosera* subgen. *Ergaleium* (Droseraceae): corms or tubers; roots or shoots? *Australian Journal of Botany, 56,* 144-152.

Cooper, C. E., Withers, P. C., Mawson, P. R., Bradshaw, S. D., Prince, J., & Roberston, H. (2002). Metabolic ecology of cockatoos in the south-west of Western Australia. *Australian Journal of Zoology, 50,* 67-76.

Cooper, C. E., Withers, P. C., Mawson, P. R., Johnstone, R., Kirkby, T., Prince, J., Bradshaw, S. D., & Robertson, H. (2003). Characteristics of Marri (*Corymbia calophylla*) fruits in relation to foraging behaviour of the Forest Red-tailed Black Cockatoo (*Calyptorhynchus banksii naso*). *Journal of the Royal Society of Western Australia, 86,* 139-142.

Cowling, R. M., & Lamont, B. B. (1985a). Seed release in *Banksia*: the role of wet-dry cycles. *Australian Journal of Ecology, 10,* 169-171.

Cowling, R. M., & Lamont, B. B. (1985b). Variation in serotiny of three *Banksia* spp. along a climatic gradient. *Australian Journal of Ecology, 10,* 345-350.

Cowling, R. M., & Lamont, B. B. (1998). On the nature of Gondwanan species flocks: diversity of Proteaceae in mediterranean Australia and South Africa. *Australian Journal of Botany, 46,* 335-355.

Cowling, R. M., Ojeda, F., Lamont, B. B., Rundel, P. W., & Lechmere-Oertel, R. (2005). Rainfall reliability, a neglected factor in explaining convergence and divergence of plant traits in fire-prone mediterranean-climate ecosystems. *Global Ecology and Biogeography, 14,* 509-519.

Cowling, R. M., Rundel, P. W., Lamont, B. B., Arroyo, M. K., & Arianoutsou, M. (1996). Plant diversity in Mediterranean-climate regions. *Trends in Ecology and Evolution, 11,* 362-366.

Cozzolino, S., & Widmer, A. (2005). Orchid diversity: an evolutionary consequence of deception? *Trends in Ecology and Evolution, 20,* 487-494.

Cramer, M. D., & Midgley, J. J. (2009). Maintenance costs of serotiny do not explain weak serotiny. *Austral Ecology, 34,* 653-662.

Craven, L. A., Edwards, R. D., & Cowley, K. J. (2014). New combinations and names in *Melaleuca* (Myrtaceae). *Taxon, 63,* 663-670.

Crisp, M. D., (1996). Convergent evolution bird-pollination in Western Australian Fabaceae, and its taxonomic implications. In S. D. Hopper, M. Harvey, J. Chappill and A. S. George (Eds), *Gondwanan heritage: Past, present and future of the Western Australian biota*, pp. 179-186. Chipping Norton: Surrey Beatty and Sons.

Crisp, M. D., Burrows, G. E., Cook, L. G., Thornhill, H., & Bowman, D. M. J. S. (2011). Flammable biomes dominated by eucalypts originated at the Cretaceous-Palaeogene boundary. *Nature Communications, 2:193 doi: 10.1038/ncomms1191*

Crisp, M. D. & Cook, L. G. (2003). Phylogeny and evolution of anomalous roots in *Daviesia* (Fabaceae: Mirbelieae). *International Journal of Plant Science, 164,* 603-612.

Crisp, M. D. & Cook, L. G. (2007). A congruent molecular signature of vicariance across multiple plant lineages. *Molecular Phylogenetics and Evolution, 43,* 1106-1117.

Crisp, M. D., Cook, L. G., & Steane, D. A. (2004). Radiation of the Australian flora: what can comparisions of molecular phylogenies across multiple taxa tell us about the evolution of diversity in present-day communities? *Philosophical Transactions of the Royal Society B, 359,* 1551-1571.

Crombie, D. S. (1992). Root depth, leaf area and daytime water relations of jarrah (*Eucalyptus marginata*) forest overstorey and understorey during summer drought. *Australian Journal of Botany, 40,* 113-122.

Crombie, D. S., Tippett, J. T., & Hill, T. C. (1988). Dawn water potential and root depth of trees and understorey species in south-western Australia. *Australian Journal of Botany, 36,* 621-631.

Crosti, R., Ladd, P. G., Dixon, K. W., & Piotto, B. (2006). Post-fire germination: the effect of smoke on seeds of selected species from the central Mediterranean basin. *Forest Ecology and Management, 221,* 306-312.

Dafni, A., & Bernhardt, P. (1990). Pollination of terrestrial orchids in southern Australia and the Mediterranean region. In M. K. Hecht, B. Wallace & R. I. Macintyre (Eds), *Evolutionary biology,* Volume 24, pp. 192-252. New York: Plenum Publishing Corporation.

Dafni A., & Calder, D. M. (1987). Pollination by deceit and floral mimesis in *Thelymitra\ antennifera* (Orchidaceae). *Plant Systematics and Evolution, 158,* 11–22.

Darnowski, D. W., Carroll, D. M., Plachno, B., Kabanoff, E., & Cinnamon, E. (2006). Evidence of protocarnivory in Triggerplants (*Stylidium* spp., Stylidiaceae). *Plant Biology, 8,* 805-812.

Dawson, T. E., & Pate, J. S. (1996). Seasonal water uptake and movement in root systems of Australian phreatophytic plants of dimorphic root morphology: a stable isotope investigation. *Oecologia, 107,* 13-20.

Day, D. A., Collins, B. G., & Rees, R. (1997). Reproductive biology of the rare and endangered *Banksia brownii* Baxter ex R.Br. (Proteaceae). *Australian Journal of Ecology, 22,* 307-315.

Day, D. A., Collins, B. G., & Rees, R. G. (1997). Reproductive biology of the rare and endangered *Banksia brownii* Baxter ex R. Br. (Proteaceae). *Australian Journal of Ecology, 22,* 307-315.

De Lange, J. H., & Boucher, C. (1990). Autoecological studies on *Audouinia capitata* (Bruniaceae). I. Plant-derived smoke as a seed germination cue. *South African Journal of Botany, 56,* 700-703.

Dell, B. (1977). Distribution and function of resins and glandular hairs in Western Australian plants. *Journal of the Royal Society of Western Australia, 59,* 119-122.

Dell, B. (1980). Structure and function of the strophiolar plug in seeds of *Albizia lophantha*. *American Journal of Botany, 67,* 556-563.

Dell, B., & Burbidge, A. H. (1981). Notes on the biology of *Pilostyles* (Rafflesiaceae) in Western Australia. *Western Australian Herbarium Research Notes, 5,* 71-79.

Dell, B., Kuo, J., & Burbidge, A.H. (1982). Anatomy of *Pilostyles hamiltonii* C.A.Gardner (Rafflesiaceae) in stems of *Daviesia*. *Australian Journal of Botany,* 30, 1-9.

Dell, B., & McComb, A. J. (1974). Resin production and glandular hairs in *Beyeria viscosa* (Labill.) Miq. (Euphorbiacae). *Australian Journal of Botany, 22,* 195-210.

Dell, B., & McComb, A. J. (1975). Glandular hairs, resin production, and habit of *Newcastelia viscida* E. Pritzel (Dicrastylidacae). *Australian Journal of Botany 22,* 195-210.

Dixon, K.W., Kuo, J., & Pate, J. S. (1983.) Storage reserves of the seed-like, aestivating organs of geophytes inhabiting granite outcrops in south-western Australia. *Australian Journal of Botany, 31,* 85-103.

Dixon, K. W., Pate, J. S., & Bailey, W. J. (1980). Nitrogen nutrition of the tuberous sundew *Drosera erythrorhiza* Lindl. with special reference to catch of arthropod fauna by its glandular leaves. *Australian Journal of Botany, 28,* 283-297.

Dixon, K. W., Sivasithamparam, K., & Read, D. J. (2002). Ericoid mycorrhizas in plant communities. In K. Sivasithamparam, K.W. Dixon & R.L. Barrett (Eds), *Microorganisms in Plant Conservation and Biodiversity,* pp. 227-239. The Hague: Kluwer Academic Publishers.

Dixon, K. W., & Tremblay, R. L. (2009). Biology and natural history of *Caladenia. Australian Journal of Botany, 57,* 247-258.

Dodd, J., Heddle, E. M., Pate, J. S., & Dixon, K. W. (1984). Rooting patterns of sandplain plants and their functional significance. In J. S. Pate and J. S. Beard (Eds), *Kwongan. Plant life of the sandplain* pp. 146-177. Nedlands, Australia: University of Western Australia Press.

Dodson, J. R., & Macphail, M. K. (2004). Palynological evidence for aridity events and vegetation change during the Middle Pliocene, a warm period in southwestern Australia. *Global and Planetary Change, 41,* 285-307.

Dolman, G. & Joseph, L. (2012). A species assemblage approach to comparative phylogeopgrahy of birds in southern Australia. *Ecology and Evolution, 2,* 354-369.

Downes, K. S., Lamont, B. B., Light, M. E., & van Staden, J. (2010). The fire ephemeral *Tersonia cyathiflora* (Gyrostemonaceae) germinates in response to smoke but not the butenolide 3-methyl-2*H*- fluro[2,3-*c*]pyran-2-one. *Annals of Botany, 106,* 381-384.

Downes, K. S., Light, M. E., Pošta, M., Kohout, L., & van Staden, J. (2013). Comparison of germination responses of *Anigozanthos flavidus* (Haemodoraceae), *Gyrostemon racemiger* and *Gyrostemon ramulosus* (Gyrostemonaceae) to smoke-water and the smoke-derived compounds karrikinolide (KAR_1) and glycernitrile. *Annals of Botany, 111,* 489-497.

Downes, K. S., Light, M. E., Pošta, M., Kohout, L., & van Staden, J. (2014). Do fire-related cues, including smoke-water, karrikinolide, glyceronitrile and nitrate, stimulate the germination of 17 *Anigozanthos* taxa and *Blancoa canescens* (Haemodoraceae)? *Australian Journal of Botany, 62,* 347-358.

Downey, P. O. (1998). An inventory of host species for each aerial mistletoe species (Loranthaceae and Viscaceae) in Australia. *Cunninghamia, 53,* 685-720.

Drake, P. L., Froend, R. H., & Franks, P. J. (2013). Smaller, faster stomata: scaling of stomata size, rate of response, and stomatal conductance. *Journal of Experimental Botany, 64,* 495-505.

Dreschler, M., Lamont, B. B., Burgman, M. A., Akçakaya, H. R., Witkowski, E. T. F., & Supriyadi, Y. (1999). Modelling the persistence of an apparently immortal *Banksia* species after fire and land clearing. *Biological Conservation, 88,* 249-259.

Duncan, D. H., Cunningham, S. A., & Nicotra, A. B. (2004). High self-pollen transfer and low fruit set in buzz-pollinated *Dianella revoluta* (Phormiaceae). *Australian Journal of Botany, 52,* 185-193.

Dundas, S. J., Fleming, P., & Hardy, G. E. StJ. (2013). Flower visitation by honey possums (*Tarsipes rostratus*) in a coastal banksia heathland infested with the plant pathogen *Phytophthora cinnamomi. Australian Mammalogy,* 35, 166-174.

Dyer, A. G., Boyd-Gery, S., McLoughlin, S., Rosa, M. G. P., Simonov, V., & Wong, B. B. M. (2012). Parallel evolution of angiosperm colour signals: common evolutionary pressures linked to hymenopteran vision. *Proceedings of the Royal Society B,* 279, 3606-3615.

Edens-Meier, R., Raguso, R. A., Westhus, E., & Bernhardt, P. (2014). Floral fraudulence: do blue *Thelymitra* species (Orchidaceae) mimic *Orthrosanthus laxus* (Iridaceae)? *Telopea, 17,* 15-28.

Edwards, R. D., Craven, L. A., Crisp, M. D. & Cook, L. G. (2010). *Melaleuca* revisited: cpDNA and morphological data confirm *Melaleuca* L. (Myrtaceae) is not monophyletic. *Taxon, 59,* 744-754.

Enright, N. J., Goldblum, D., Ata, P., & Ashton, D. H. (1997). The independent effects of heat, smoke and ash on emergence of seedlings from the soil seed bank of a healthy *Eucalyptus* woodland in Grampians (Gariwerd) National Park, western Victoria. *Australian Journal of Ecology,* 22, 81-88

Enright, N. J., & Lamont, B. B. (1989). Fire temperatures and follicle-opening requirements in ten *Banksia* species. *Australian Journal of Ecology, 14,* 107-113.

Enright, N.J., & Lamont, B. B. (1992). Survival, growth and water relations of *Banksia* seedlings in a sand mine rehabilitation site and adjacent scrub-heath sites. *Journal of Applied Ecology 29,* 663-671.

Enright, N. J., Marsula, R., Lamont, B. B., & Wissel, C. (1998). The ecological significance of canopy seed storage in fire-prone environments: a model for nonsprouting shrubs. *Journal of Ecology, 86,* 946-959.

Enright, N. J., Mosner, E., Miller, B. P., Johnson, N., & Lamont, B. B. (2007). Soil vs canopy seed storage and plant species coexistence in species-rich Australian shrublands. *Ecology, 88,* 2292-2304.

Esther, A., Groeneveld, G., Enright, N. J., Miller, B. P., Lamont, B. B., Perry, G. L. W., Blank, F. B., & Jeltsch, F. (2010). Sensitivity of plant functional types to climate change: classification tree analysis of a simulation model. *Journal of Vegetation Science, 21,* 447-461.

Esther, A., Groeneveld, J., Enright, N.J., Miller, B.P., Lamont, B.B., Perry, G.L., Tietjen, B., Jeltsch, F. (2011). Low-dimensional trade-offs fail to explain richness and structure in species-rich plant communities. *Theoretical Ecology, 4,*495-511.

Exley, E. M., (1998). New *Euryglossa* (*Euhesma*) bees (Hymenoptera: Colletidae: Euryglossinae) associated with the Australian plant genus *Eremophila* (Myoporaceae). *Records of the Western Australian Museum, 18,* 419-437.

Exley, E. M. (2004). Revision of the genus *Dasyhesma* Michener (Apoidea: Colletidae: Euryglossinae). *Records of the Western Australian Museum, 22,* 129-146.

Farrington, P., Greenwood, E. A. N., Bartle, J. D., & Watson, G. D. (1989). Evaporation from a *Banksia* woodland on a groundwater mound. *Journal of Hydrology 105,* 173-186.

Findlay, G. P. & Findlay, N. (1975). Anatomy and movement of the column in *Stylidium. Australian Journal of Plant Physiology*, *2,* 597-621.

Findlay, N. & Findlay, G. P. (1984). Movement of potassium ions in the motor tissue of *Stylidium. Australian Journal of Plant Physiology, 11,* 451-457.

Findlay, N. & Findlay, G. P. (1989). The structure of the column in *Stylidium. Australian Journal of Botany, 37,* 81-101.

Flematti, G. R., Ghisalberti, E. L., Dixon, K. W., & Trengrove, R. D. (2004). A compound from smoke that promotes seed germination. *Science, 305,* 977.

Flematti, G. R., Merritt, D. J., Piggott, M. J., Trengrove, R. D. Smith, S. M., Dixon, K. W., & Ghisalberti, E. L. (2011). Burning vegetation produces cyanohydrins that liberate cyanide and stimulate seed germination. *Nature Communications,* doi: 10.1038/ncomms1356

Fleming, P. A., Anderson, H., Prendergast, A. S., Bretz, M. R., Valentine, L. E., & Hardy, G. E. StJ. (2014). Is the loss of Australian digging mammals contributing to a deterioration in ecosystem function? *Mammal Review, 44,* 95-108.

Forster, M. A., & Bonser, S. P. (2009). Heteroblastic development and the optimal partitioning of traits among contrasting environments in *Acacia implexa. Annals of Botany, 103,* 95-105.

Gaff, D. F., & Oliver, M. (2013). The evolution of desiccation tolerance in angiosperm plants: a rare yet common phenomenon. *Functional Plant Ecology, 40,* 315-328.

Gaff, D. F., Zee, S. Y., & O'Brien, T. P. (1976). The fine structure of dehydrated and reviving leaves of *Borya nitida* Labill. - a desiccation-tolerant plant. *Australian Journal of Botany, 24,* 225–236.

Garavanta, C. A. M., Wooller, R. D., & Richardson, K. C. (2000). Movement patterns of honey possums, *Tarsipes rostratus*, in the Fitzgerald River National Park, Western Australia. *Wildlife Research, 27,* 179-183.

Gaskett, A. C. (2012). Floral shape mimicry and variation in sexually deceptive orchids with a shared pollinator. *Biological Journal of the Linnean Society, 106,* 469-481.

Gaskett, A. C. & Herberstein, M. E. (2010). Colour mimicry and sexual deception by tongue orchids (*Cryptostylis*). *Naturwissenschaften*, 97, 97-102.

Gaskett, A. C., Winnick, C. G., & Herberstein, M. E. (2008). Orchid sexual deceit provokes ejaculation. *American Naturalist. 171,* E206–E212.

George, A. S. (2002a). The south-western Australian flora in autumn: 2001 Presidential Address. *Journal of the Royal Society of Western Australia, 85,* 1-15.

George, A. S. (2002b). *The long dry. Bush colours of summer and autumn in south-western Australia.* Kardinya, Western Australia: Four Gables Press.

George, A. S., & Barrett, M. D. (2010). Two new taxa of *Verticordia* (Myrtaceae: Chamelaucieae) from south-western Australia. *Nuytsia, 20,* 309-318.

George, E. S., & Pieroni, M. (2002). *Verticordia: the turner of hearts. Crawley,* Western Australia: University of Western Australia Press.

Gibson, N., Meissner, R., Markey, A. S., & Thompson, W. A. (2012). Patterns of plant diversity in ironstone ranges in arid south-western Australia. *Journal of Arid Environments, 77,* 25-31.

Gibson, N., Yates, C. J., & Dillon, R. (2010). Plant communities of the ironstone ranges of south western Australia: hotspots for plant diversity and mineral deposits. *Biodiversity and Conservation, 19,* 3951-3962.

Gibson, T. C., & Waller, D. M. (2009). Evolving Darwin's 'most wonderful' plant: ecological steps to a snap-trap. *New Phytologist, 183,* 575-587.

Gill, A. M. (1975). Fire and the Australian flora – a review. *Australian Forester, 38,* 4-25.

Glassford, D. K., & Semeniuk, V. (1995). Desert-aeolian origin of late Cenozoic regolith in arid and semi-arid Southwestern Australia. *Palaeogeography, Palaeoclimatology, Palaeoecology, 114,* 131-166.

Gove, A. D., Majer, J. D., & Dunn, R. R. (2007). A keystone ant species promotes seed dispersal in a 'diffuse' mutualism. *Oecologia, 153,* 687-697.

Groom, P. K. (2002). Seedling water stress response of two sandplain *Banksia* species differing in ability to tolerate drought. *Journal of Mediterranean Ecology, 3,* 3-9.

Groom, P. K. (2003). Groundwater-dependency and water relations of four Myrtaceous shrub species during a prolonged summer drought. *Journal of the Royal Society of Western Australia, 86,* 31-40.

Groom, P. K. (2004). Rooting depth and plant water relations explain species distribution patterns within a sandplain landscape. *Functional Plant Biology, 31,* 423-428.

Groom, P. K., Froend, R. H., & Mattiske, E. M. (2000). Impact of groundwater abstraction on a *Banksia* woodland, Swan Coastal Plain, Western Australia. *Ecological Management and Restoration,* 1, 117-124.

Groom, P. K., Froend, R. H., Mattiske, E. M., & Koch, B.L. (2000). Myrtaceous shrub species respond to long-term decreasing groundwater levels on the Gnangara Groundwater Mound, northern Swan Coastal Plain. *Journal of the Royal Society of Western Australia, 83,* 75-82.

Groom, P. K., Froend, R. H., & Mattiske, E. M. (2001). Long-term changes in vigour and distribution of *Banksia* and *Melaleuca* overstorey species on the Swan Coastal Plain. *Journal of the Royal Society of Western Australia, 84,* 63-69.

Groom, P. K., & Lamont, B.B. (1996a). Ecogeographcical analysis of *Hakea* in south-western Australia, with special reference to leaf morphology and life form. *Australian Journal of Botany, 44*, 527-542.

Groom, P. K., & Lamont, B.B. (1996b). Reproductive ecology of non-sprouting and resprouting *Hakea* species (Proteaceae) in southwestern Australia. In S. D. Hopper, M. Harvey, J. Chappill and A. S. George (Eds), *Gondwanan heritage: Past, present and future of the Western Australian biota*, pp. 239-248. Chipping Norton: Surrey Beatty and Sons.

Groom, P. K., & Lamont, B. B. (1999). Which common indices of sclerophylly best reflect differences in leaf structure? *Écoscience, 6*, 471-474.

Groom, P. K., & Lamont, B. B. (2007). Fruit-seed relations in *Hakea*: serotinous species invest more dry matter in predispersal seed protection. *Australian Journal of Ecology, 22*, 352-355.

Groom, P. K., & Lamont, B.B. (2010). Seed phosphorus accumulation in Proteaceae seeds: a synthesis. *Plant and Soil, 334*, 61-72.

Groom, P. K., & Lamont, B. B. (2011). Regional and local effects on reproductive allocation in epicormic and lignotuberous populations of *Banksia menziesii*. *Plant Ecology, 212*, 2003-2011.

Groom, P. K., Lamont, B. B., & Duff, H. C. (1994). Self-crypsis in *Hakea trifurcata* as an avian granivore deterrent. *Functional Ecology, 8*, 110-117.

Groom, P. K., Lamont, B. B., & Kupsky, L. (1994). Contrasting morphology and ecophsyiology of co-occurring broad and terete leaves in *Hakea trifurcata* (Proteaceae) *Australian Journal of Botany, 42*, 307-320.

Groom, P. K., Lamont, B. B., Leighton, S., Leighton, P., & Burrows, C. (2004). Heat damage in sclerophylls is influenced by their leaf properties and plant environment. *Écoscience,* 11, 94-101.

Groom, P. K., Lamont, B. B., & Markey, A. S. (1997). Influence of leaf type and plant age on leaf structure and sclerophylly in *Hakea* (Proteaceae). *Australian Journal of Botany, 45*, 827-838.

Grosso, B., Saint-Martin, M., & Vassal, J. (1994). Stomatal types of the genus *Acacia* (Fabaceae, Mimosoideae): an appraisal of diversity and taxonomic interest. *Botanical Journal of the Linnaean Society, 116*, 325-341.

Halliday, J., & Pate, J. S. (1976). Symbiotic nitrogen fixation by coralloid roots of the cycad *Macrozamia riedlei*: physiological characteristics and ecological significance. *Australian Journal of Plant Physiology, 3*, 349-358.

Hanley, M. E., & Lamont, B. B. (2000). Heat pre-treatment and the germination of soil- and canopy-stored seeds of south-western Australia. *Acta Oecologica, 21*, 315-321.

Hanley, M. E., Lamont, B. B., & Armbruster, W.S. (2009). Pollination and plant defence traits co-vary in Western Australia hakeas. *New Phytologist, 182*, 251-260.

Hanley, M. E., Lamont, B. B., Fairbanks, M. M., & Rafferty, C. M. (2007). Plant structural traits and their role in anti-herbivory defence. *Perspectives in Plant Ecology, Evolution and Systematics, 8*, 157-178.

Hassiotou, F., Evans, J. R., Ludwig, M., & Veneklaas, E. J. (2009). Stomatal crypts may facilitate diffusion of CO_2 to adaxial mesophyll cells in thick sclerophylls. *Plant, Cell and Environment, 32*, 1596-1611.

Hawkeswood, T. J. (1992). Observations on insect pollination of *Melaleuca lanceolata* Otto and *Verticordia picta* Endl. (Myrtaceae), *Leucopogon parviflorus* (Andr.) Lindl. (Epacridaceae) and *Eryngium pinnatifidum* Bunge (Apiaceae) in Western Australia. *Giornale Italiano di Entomologia, 6*, 239-242.

Hawkeswood, T. J. (1993). Some notes on native bees (Hymenoptera: Medachilidae: *Chalicodoma*) visiting the flowers of *Calytrix fraseri* A. Cunn. (Myrtaceae) in Western Australia. *Giornale Italiano di Entomologia, 6*, 235-237.

He, T. (2014). Ecological divergence and evolutionary transition of resprouting types in *Banksia attenuata*. *Ecology and Evolution, 43*, 235-248.

He, T., Krauss, S., Lamont, B. B., Miller, B. P., & Enright, N. J. (2004). Long-distance seed dispersal in a metapopulation of *Banksia hookeriana* inferred by population allocation analysis from AFLP data. *Molecular Ecology, 13,* 1099-1109.

He, T., Lamont, B. B., & Downes, K. S. (2011). Banksia born to burn. *New Phytologist, 191,* 184-196.

He, T., Lamont, B., Enright, N., Krauss, S., & Merwin, L. (2011). Migration potential as a new predictor of long-distance dispersal rate for plants. *Nature Proceedings,* <http://dx.doi.org/10.1038/npre.2011.5890.1>

He, T., Lamont, B. B., Krauss, S. L., Enright, N. J., Miller, B. P., & Gove, A. D. (2009). Ants cannot account for interpopulation dispersal of the arillate pea *Daviesia triflora. New Phytologist, 181,* 725-733.

He, T., Lamont, B. B., Krauss, S. L., Enright, N. J., & Miller, B. P. (2009). Long-distance dispersal of seeds in the fire-tolerant shrub *Banksia attenuata. Ecography, 32,* 571-580.

He, T., Pausas, J. G., Belcher, C. M., Schwilk, D. W., & Lamont, B. B. (2012). Fire-adapted traits of *Pinus* arose in the fiery Cretaceous. *New Phytologist 194,* 751-759.

Heide-Jorgensen, H. S. (1991). Anatomy and ultrastructure of the haustorium of *Cassytha pubescens* R.Br. I. The adhesive disc. *Botanical Gazette, 142,* 321-334.

Herath, D. N., Lamont, B. B, Enright, N. J., & Miller, B. P. (2009). Comparison of postmine rehabilitated and natural shrubland communities in southwestern Australia. *Restoration Ecology, 17,* 577–585.

Hidayati, S. N., Walck J. L., Merritt, D. J., Turner, S. R., Turner, D. W., & Dixon, K. W. (2012). Sympatric species of *Hibbertia* (Dilleniaceae) vary in dormancy break and germination requirements: implications for classifying morphophysiological dormancy in Mediterranean biomes. *Annals of Botany, 109,* 1111-1123.

Hnatiuk R. J., & Hopkins, A. J. M. (1980). Western Australian species-rich kwongan (sclerophyllous shrubland) affected by drought. *Australian Journal of Botany*, 28, 573-585.

Hocking, P. J. (1981). Accumulation of mineral nutrients by developing fruits of prickly plume grevillea (*Grevillea annulifera* F. Muell.). *Australian Journal of Botany, 29,* 507-520.

Hocking, P. J. (1986). Mineral nutrient composition of leaves and fruit of selected species of *Grevillea* from south-western Australia, with special reference to *Grevillea leucopteris* Meissn. *Australian Journal of Botany, 34,* 155-164.

Hopper, S. D. (1979). Biogeographical aspects of speciation in the south-west Australian flora. *Annual Review of Ecology and Systematics, 10,* 399-422.

Hopper, S. D. (1980). Bird and mammal pollen vectors in *Banksia* communities at Cheyne Beach, Western Australia. *Australian Journal of Botany, 28,* 61-75.

Hopper, S. D. (1981). Honeyeaters and their winter food plants on granite rocks in the central wheatbelt of Western Australia. *Australian Wildlife Research, 8,* 187-197.

Hopper, S. D. (2009). Taxonomic turmoil down-under: recent developments in Australian orchid systematics. *Annals of Botany, 104,* 447-455.

Hopper, S. D. (2009). OCBIL theory: towards an integrated understanding of the evolution, ecology and conservation of biodiversity on old, climatically buffered, infertile landscapes. *Plant and Soil, 322,* 49-86.

Hopper, S. D. (2010). *Nuytsia floribunda*. Loranthaceae. *Curtis's Botanical Magazine, 26,* 333-368.

Hopper, S. D., & Brown, A. P. (2004). Robert Brown's *Caladenia* revisited, including a revision of its sister genera *Cyanicula, Ericksonella* and *Pheladenia* (Caladeniinae: Orchidaceae). *Australian Systematic Botany, 17,* 171-240.

Hopper, S. D., & Brown, A. P. (2006). Australia's wasp-pollinated flying duck orchids revised (*Paracaleana*: Orchidaceae). *Australian Systematic Botany, 19,* 211-244.

Hopper, S. D., & Brown, A. P. (2007). A revision of Australia's hammer orchards (*Drakaea:* Orchidaceae) with some field data on species-specific sexually deceived wasp pollinators. *Australian Systematic Botany, 20,* 252-285.

Hopper, S. D., & Burbidge, A. H. (1978). Assortative pollination by red wattlebirds in a hybrid population of *Anigozanthos* Labill. (Haemodoraceae). *Australian Journal of Botany, 26,* 335-350.

Hopper, S. D., & Gioia, P. (2004). The southwestern Australian floristic region: evolution and conservation of a global hot spot of biodiversity. *Annual Review of Ecology, Evolution and Systematics, 35,* 623-650.

Hopper, S. D., & Moran, G. F. (1981). Bird pollination and the mating system of *Eucalyptus stoatei. Australian Journal of Botany, 29,* 625-638.

Hopper, S. D., Smith, R. J., Fay, M. F., Manning, J. C., & Chase, M. W. (2009). Molecular phylogenetics of Haemodoraceae in the Greater Cape and Southwest Australian Floristic Region. *Molecular Phylgenetics and Evolution,* 51, 19-30.

Hos, D., (1975). Preliminary investigation of the palynology of the upper Eocene Werillup Formation, Western Australia. *Journal of the Royal Society of Western Australia, 58,* 1–14.

Horton, J. L., & Hart, S. C. (1998). Hydraulic lift: a potentially important ecosystem process. *Trends in Ecology and Evolution, 13,* 232-235.

Houston, T. F., (1983). An extraordinary new bee and adaptation of palpi for nectar-feeding in some Australian Colletidae and Pergidae (Hymenoptera). *Journal of the Australian Entomological Society, 22,* 263-270.

Houston, T. F., (1989). *Leioproctus* bees associated with Western Australian smoke bushes (*Conospermum* spp.) and their adaptations for foraging and concealment (Hymenoptera: Colletidae: Paracoletini). *Records of the Western Australian Museum, 14,* 275-292.

Houston, T. F., (1991). Two new and unusual species of the bee genus *Leioproctus* Smith (Hymenoptera: Collectidae), with notes on their behaviour. *Records of the Western Australian Museum, 15,* 83-96.

Houston, T. F., (1992). Three new, monolectic species of *Euryglossa* (*Euhesma*) from Western Australia (Hymenoptera: Colletidae). *Records of the Western Australian Museum, 15,* 719-728.

Houston, T. F., & Ladd, P. G. (2002). Buzz pollination in the Epacridaceae. *Australian Journal of Botany, 50,* 83-91.

Houston, T. F., Lamont, B. B., Radford, S., & Errington, S. G. (1993). Apparent mutualism between *Verticordia nitens* and *V. aurea* (Myrtaceae) and their oil-ingesting bee pollinators (Hymenoptera, Colletidae). *Australain Journal of Botany, 41,* 369-380.

Howell, G. J., Slater, A. T., & Knox, R. B. (1993). Secondary pollen presentation in angiosperms and its biological significance. *Australian Journal of Botany, 41,* 417-438.

Hughes, L., & Westoby, M. (1992). Fate of seeds adapted for dispersal by ants in Australian sclerophyll vegetation. *Ecology, 73,* 1285-1299.

Hynson, N. A., Madsen, T. P., Selosse, A-A, Adam, I. K. U., Ogura-Tsujita, Y., Roy, M., & Gebauer, G. (2013). The physiological ecology of mycoheterotrophy. In V. S. F. T. Merckx (ed.) *Mycoheterotrophy: The biology of plants living on fungi*, pp. 297-342. New York: Springer.

Indsto, J. O., Weston, P. H., Clements, M. A., Dyer, A. G., Batket, M., & Whelan, R. (2006). Pollination of *Diuris maculata* (Orchidaceae) by male *Trichocolletes venustus* bees. *Australian Journal of Botany, 54,* 669-679.

Itzstein-Davey, F. (2004). A spatial and temporal Eocene palaeoenvironmental study, focusing on the Proteaceae family, from Kambalda, Western Australia. *Review of Palaeobotany and Palynology, 131,* 159-180.

Itzstein-Davey, F. (2007). Changes in the abundance and diversity of Proteaceae in south-western Australia: a review of an integrated palaeoenvironmental study. *Geographical Research,* 45, 43-53.

Jersáková, J. Johnson, S. D., & Kindlmann, P. (2006). Mechanisms and evolution of deceptive pollination in orchids. *Biological Reviews, 81,* 219-235.

Johnson, K. A. (2012). Pollination ecology and evolution of epacrids. PhD thesis, University of Tasmania. Australia.

Johnson, K. A. (2013). Are there pollination syndromes in the Australian epacrids (Ericaceae: Styphelioideae)? A novel statistical method to identify key floral traits per syndrome. *Annals of Botany, 112,* 141-149.

Johnson, S. D., Hobbhahn, N., & Bytebier, B. (2013). Ancestral deceit and labile evolution of nectar production in the African orchid genus *Disa. Biological Letters, 9,* 20130500

Jordan, G. J., Brodribb, T. J., Blackman, C. J, & Weston, P. H. (2013). Climate drives vein anatomy in Proteaceae. *American Journal of Botany, 100,* 1483-1493.

Jordan, G. J., Dillon, R. A., & Weston, P. H. (2005). Solar radiation as a factor in the evolution of scleromorphic leaf anatomy in Proteaceae. *American Journal of Botany, 92,* 789–796.

Jordan, G. J., Weston, P. H., Carpenter, R. J., Dillon, R. A., & Brodribb, T. J. (2008). The evolutionary relations of sunken, covered and encrypted stomata to dry habitats in Proteaceae. *American Journal of Botany, 95,* 521-530.

Joseph, L., Toon, A., Nyári, A. S., Longmore, N. W., Rowe, K. M. C., Haryoko, T., Trueman, J., & Gardner, J. L. (2014) A new synthesis of the molecular systematics and biogeography of honeyeaters (Passeriformes: Meliphagidae) highlights biogeographical complexity of a spectacular avian radiation. *Zoologica Scripta,* (in press)

Kaplan, D. R., (1980). Heteroblastic leaf development in *Acacia* – morphological and morphogenetic implications. *La Cellule, 73,* 135-203.

Keeley, J. E., & Fotheringham, C. J. (1998). Smoke-induced seed germination in California chaparral. *Ecology, 79,* 2320–2336.

Keeley, J. E., Pausas, J. G., Rundel, P. W., Bond, W. J., & Bradstock, R. A. (2011). Fire as an evolutionary pressure shaping plant traits. *Trends in Plant Science, 16,* 406-411.

Keighery, G. J., (1975). Parallel evolution of floral structures in *Darwinia* (Myrtaceae) and *Pimelea* (Thymelaeaceae). *Western Australian Naturalist, 13,* 46-50.

Keighery, G. J. (1980). Bird pollination in south western Australia: A checklist. *Plant Systematics and Evolution, 135,* 171-176.

Keighery, G. J. (1981). The breeding system of *Emblingia* (Emblingiaceae). *Plant Systematics and Evolution, 137,* 63-65.

Keighery, G. J. (1996). Phytogeography, biology and conservation of Western Australian Epacridaceae. *Annals of Botany, 77,* 347-355.

Keppel, G., van Niel, K. P., Wardell-Johnson, G. W., Yates, C. J., Byrne, M., Mucina, L., Schut, A. G. T., Hopper, S. D., & Franklin, S. E. (2012). Refugia: identifying and understanding safe havens for biodiversity under climate change. *Global Ecology and Biogeography, 21,* 393-404.

Kim, J., Walck, J. L., Hidayati, S. N., Merritt, D. J., & Dixon, K. W. (2009). Germinability of seeds stored in capsules on plants of two myrtaceous shrubs: differences among age cohorts and between species. *Australian Journal of Botany, 57,* 495-501.

Knight, C. A., & Ackerly, D. D. (2002). An ecological and evolutionary analysis of photosynthetic thermotolerance using the temperature-dependent increase in fluorescence. *Oecologia, 130,* 505-514.

Knight, C. A., & Ackerly, D. D. (2003). Evolution and plasticity of photosynthetic thermal tolerance, specific leaf area and leaf size: congeneric species from desert and coastal environments. *New Phytologist 160,* 337-347.

Knox, R. B., Kendrick, J., Bernhardt, P., Marginson, R., Bresford, G., Baker, I., & Baker, H. G. (1985). Extrafloral nectaries as adaptations for bird pollination in *Acacia terminalis. American Journal of Botany, 72,* 11985-1196.

Kuo J., Hocking, P.J., Pate, J. S. (1982). Nutrient reserves in seeds of selected proteaceous species from south-western Australia. Australian Journal of Botany, *30,* 231-249.

Lach, L., Hobbs, R.J., & Majer, J. D. (2009). Herbivory-induced extrafloral nectar increases native and invasive ant worker survival. *Population Ecology,* 51, 237-243.

Ladd, P. G., Alkema, A.J., & Thomson, G. J. (1996). Pollen presenter morphology and anatomy in *Banksia* and *Dryandra*. *Australian Journal of Botany, 44*, 447-471.

Ladd, P. G., & Connell, S. W. (1994). Andromonoecy and fruit set in three genera of the Proteaceae. *Botanical Journal of the Linnaean Society, 116*, 77-88.

Ladd, P. G., & Enright, N. J. (2011). Ecology of fire-tolerant Podocarps in temperate Australian forests. *Smithsonian Contributions to Botany, 95*, 141-155.

Ladd, P. G., Parnell, J. A. N., & Thomson, G. (1999). Anther diversity and function in *Verticordia* DC. (Myrtaceae). *Plant Systematics and Evolution, 219*, 79-97.

Lambers, H., Brundrett, M. C., Raven, J. A., & Hopper, S. D. (2010). Plant mineral nutrition in ancient landscapes: high plant species diversity on infertile soils is linked to functional diversity for nutritional strategies. *Plant and Soil, 334*, 11-31.

Lambers, H., Raven, J. A., Shaver, G. R., & Smith, S. E. (1998). Plant nutrient-acquisition strategies change with soil age. *Trends in Ecology and Evolution*, 23, 95-103.

Lambers, H., Raven, J. A., Shaver, g. R., & Smith, S. E. (2008). Plant nutrient-acquistion strategies change with soil age. *Trends in Ecology and Evolution, 23*, 95-103.

Lambers, H., Shane, M. W., Cramer, M. D., Pearse, S. J., & Veneklass, E. J. (2006). Root structure and functioning for efficient acquisition of phosphoros: matching morphological and physiological traits. *Annals of Botany, 98*, 693-713.

Lamont, B. (1974). The biology of dauciform roots in the sedge *Cyathochaete avenacea*. *New Phytologist, 73*, 985-996.

Lamont, B. (1979). Extrafloral nectaries in Australian plants, with special reference to *Acacia*. *Annual Report Mulga Research Centre, 2*, 15-18.

Lamont, B. (1981a). Morphometrics of the aerial roots of *Kingia australis* (Liliales). *Australian Journal of Botany*, 29, 10-25.

Lamont, B. (1981b). Availability of water and inorganic nutrients in the persistent leaf bases of the grass tree *Kingia australis* and the uptake and translocation of labelled phosphate by the embedded aerial roots. *Physiologia Plantarum*, 52, 181-186.

Lamont, B. (1982a). Gas content of berries of the Australian mistletoe *Amyema preissii* and the effect of maturity, viscin, temperature and carbon dioxide on germination. *Journal of Experimental Botany, 135*, 790-798.

Lamont, B. B. (1982b). Mechanisms for enhancing nutrient uptake in plants, with particular reference to Mediterranean South Africa and Western Australia. *Botanical Review, 48*, 597-689.

Lamont, B. (1983). Germination of mistletoes. In Calder, D. M. and Bernhardt, P. (Eds.), *The biology of mistletoes*. pp. 129-143. Sydney, Australia: Academic Press.

Lamont, B. (1984a). Specialized modes of nutrition. In: Pate, J.S. and Beard, J.S. (Eds.). *Kwongan: plant life of the sandplain*. pp. 126-145. Crawley, University of Western Australian Press.

Lamont, B. (Ed.) (1984b). Natural resources of the Avondale Flora and Fauna Reserve, Beverley, Western Australia. Curtin School of Biology Bulletin No. 6.

Lamont, B. (1985a). The significance of flower colour change in eight co-occurring shrub species. *Botanical Journal of the Linnaean Society, 90*, 145-155.

Lamont, B. (1985b). Dispersal of the winged fruits of *Nuytsia floribunda* (Loranthaceae). *Australian Journal of Ecology, 10*, 187-193.

Lamont, B. (1982c). The reproductive biology of *Grevillea leucopteris* (Proteaceae), including reference to its glandular hairs and colonizing potential. *Flora, 172*, 1-20.

Lamont, B. B. (1988). Sexual versus vegetative reproduction in *Banksia elegens*. *Botanical Gazette, 149*, 370-375.

Lamont, B. B. (1989). Biotic and abiotic interactions in *Banksia* woodland. *Journal of the Royal Society of Western Australia, 71*, 99-100.

Lamont, B. B. (1991). Canopy seed storage and release: what's in a name? *Oikos, 60*, 266-268.

Lamont, B. B. (1992). Functional interactions within plants - the contribution of keystone and other species to biological diversity. In R. J. Hobbs, R.J. (Ed.), *Biodiversity of mediterranean ecosystems in Australia.* pp. 95-127. Chipping Norton: Surrey Beatty and Sons.
Lamont, B. B. (1994). Mineral nutrient relations in mediterranean regions of California, Chile and Australia. In M. T. Kalin-Arroyo, P, Zedler &M. Fox (Eds), *Ecology and biogeography of mediterranean ecosystems in Chile, California and Australia.*vpp. 211-235, New York: Springer-Verlag.
Lamont, B. B. (1995). Testing the effect of ecosystem composition/structure on its functioning. *Oikos, 74,* 283-295.
Lamont, B.B. (1996). Conservation biology of banksias in southwestern Australia. In S. D. Hopper, M. Harvey, J. Chappill and A. S. George (Eds), *Gondwanan heritage: Past, present and future of the Western Australian biota*, pp. 292-298. Chipping Norton: Surrey Beatty and Sons.
Lamont, B. B. (2003). Structure, ecology and physiology of root clusters—a review. *Plant and Soil, 248,* 1-19.
Lamont, B. B., & Barker, M. J. (1988). Seed bank dynamics of a serotinous, fire-sensitive *Banksia* species. *Australian Journal of Botany, 36,* 193-203.
Lamont, B. B., & Bergl, S. M. (1991). Water relations, shoot and root architecture, and phenology of three co-occurring *Banksia* species: no evidence for niche differentiation in the pattern of water use. *Oikos, 60,* 291-298.
Lamont, B. B., & Barrett, G. J. (1988). Constraints on seed production and storage in a root-suckering *Banksia*. *Journal of Ecology, 76,* 1069–1082.
Lamont, B. B., & Collins, B. G. (1988). Flower colour change in *Banksia ilicifolia*: a signal for pollinators. *Australian Journal of Ecology, 13,* 129-135.
Lamont, B. B., & Connell, S. W. (1996). Biogeography of *Banksia* in southwestern Australia. *Journal of Biogeography, 23,* 295-309.
Lamont, B. B., & Cowling, R. M. (1984). Flammable infructescences in *Banksia*: a fruit-opening mechanism. *Australian Journal of Ecology, 9,* 295-296.
Lamont, B. B., & Downes, K. S. (2011). Fire-stimulated flowering among resprouters and geophytes in Australia and South Africa. *Plant Ecology, 212,* 2111-2125.
Lamont, B. B. & Downes, S. (1979). The flowering and fire history of the grass trees *Xanthorrhoea preissii* and *Kingia australis*. *Journal of Applied Ecology,* 16, 893-899.
Lamont, B. B., Downes, S., & Fox, J. E. D. (1977). Importance-value curves and diversity indices applied to a species-rich heathland in Western Australia. *Nature*, 265, 438-441.
Lamont, B. B., & Enright, N. J. (2000). Adaptive advantages of aerial seed banks. *Plant Species Biology, 15,* 157-166.
Lamont, B. B., Enright, N. J., & He, T. (2011). Fitness and evolution of resprouters in relation to fire. *Plant Ecology, 212,* 1945-1957.
Lamont, B. B., & Groom, P. K. (1998). Seed and seedling biology of woody fruited Proteaceae. *Australian Journal of Botany, 46,* 387-406.
Lamont, B. B., & Groom, P. K. (2002). Green cotyledons of two *Hakea* species control seedling mass and morphology by supplying mineral nutrients rather than organic compounds. *New Phytologist, 153,* 101-110.
Lamont, B. B., & Groom, P. K. (2013). Seeds as a source of carbon, nitrogen and phosphorus for seedling establishment in temperate regions: a synthesis. *American Journal of Plant Sciences,* 4, 30-40.
Lamont, B. B., Groom, P. K., & Cowling, R. M. (2002). High leaf mass per area of related species assemblages may reflect low rainfall and carbon isotope discrimination rather than low phosphorus and nitrogen concentrations. *Functional Ecology, 16,* 403-412.

Lamont, B. B., Groom, P. K., Richards, M. B., & Witkowski, E. T. F. (1999). Recovery of communities after fire—the role of species identity and functional attributes. *Diversity and Distributions, 5,* 15-26.

Lamont, B. B., & He. T. (2012). Fire-adapted Gondwanan Angiosperm floras evolved in the Cretaceous. *BMC Evolutionary Biology, 12,* 223.

Lamont, B. B., He, T., & Downes, K. S. (2013). Adaptive responses to directional trait selection in the Miocene enabled Cape proteas to colonize the savanna grasslands. *Evolutionary Ecology, 27,* 1099-1115.

Lamont, B. B., He, T., Enright. N. J., Krauss. S. L., & Miller, B.P. (2003). Anthropogenic disturbance promotes hybridization between *Banksia* species by altering their biology. *Journal of Evolutionary Biology, 16,* 551-557.

Lamont, B. B., Hopkins, A. J. M., & Hnatiuk, R. J. (1984). The flora - composition, diversity and origins. In J. S. Pate and J. S. Beard (Eds), *Kwongan. Plant life of the sandplain.* pp. 27-50. Nedlands, Australia: University of Western Australia -Press.

Lamont, B. B., & Lamont, H.C. (2000). Utilizable water in leaves of eight arid species as derived from pressure-volume curves and chlorophyll fluorescence. *Physiologia Plantarum, 110,* 64-71.

Lamont, B. B., le Maitre, D. C., Cowling, R.M., & Enright, N. J. (1991). Canopy seed storage in woody plants. *The Botanical Review, 57,* 277-317.

Lamont, B. B., & Markey, A. (1995). Biogeography of fire-killed and resprouting *Banksia* species in south-western Australia. *Australian Journal of Botany, 43,* 283-303.

Lamont, B., & Perry, M. (1977). The effects of light, osmotic potential and atmospheric gases on germination of the mistletoe *Amyema preissii*. *Annals of Botany, 41,* 203-209.

Lamont, B. B, Pérez-Fernández, M., & Rodríguez-Sánchez, J. (2015). Soil bacteria hold the key to root cluster formation. *New Phytologist,* 206, 1156-1162, doi: 10.1111/nph.13228.

Lamont, B. B., Ralph, C. S., & Christensen, P. (1985). Mycophagous marsupials as dispersal agents for ectomycorrhizal fungi on *Eucalyptus calophylla* and *Gastrolobium bilobum*. *New Phytologist,* 101, 651-656.

Lamont, B. B., & Ryan, R.A. (1977). The formation of coralloid roots by cycads under sterile conditions. *Phytomorphology, 27,* 426-29.

Lamont, B. B., & Runciman, H. V. (1993). Fire may stimulate flowering, branching, seed production and seedling establishment in two kangaroo paws (Haemodoraceae). *Journal of Applied Ecology,* 30, 256-264.

Lamont, B. B., Witkowski, E. T. F., & Enright, N. J. (1993). Post-fire litter microsites: safe for seeds, unsafe for seedlings. *Ecology, 74,* 501-512.

Lamont, B. B., Wittkuhn, R., & Korczynskyj, D. (2004). Ecology and ecophysiology of grasstrees. *Australian Journal of Botany, 52,* 561-582.

Lawes, M. J., Richards, A., Dathe, J., & Midgley, J. J. (2011). Bark thickness determines fire resistance of selected tree species from fire-prone tropical savanna in north Australia. *Plant Ecology, 212,* 2057-2069.

Leigh, A., Sevanto, S., Ball, M. C., Close, J. D., Ellsworth, D. S., Knight, C. A., Nicotra, A. B., & Vogel, S. (2012). Do thick leaves avoid thermal damage in critically low wind speeds? *New Phytologist, 194,* 477-487.

Lengyel, S., Gove, A. D., Latimer, A. M., Majer, J. D., & Dunn, R. R. (209). Ants sow the seeds of global diversification in flowering plants. *PLos ONE, 4,* e5480.

Lengyel, S., Gove, A. D., Latimer, A. M., Majer, J. D., & Dunn, R. R. (2010). Convergent evolution of seed dispersed by ants, and phylogeny and biogeography in flowering plants: a global survey. *Perspectives in Plant Ecology, Evolution and Systematics,* 12, 43-55.

Light, M. E., Daws, M. I., & van Staden, J. V. (2009). Smoke-derived butenolide: towards understanding its biological effects. *South African Journal of Botany, 75,* 1-7.

Loveys, B. R., Loveys, B. R., & Tyerman, S.D. (2001). Water relations and gas exchange of the root hemiparasite *Santalum acuminatum* (quandong). *Australian Journal of Botany, 49,* 479-486.

Lowrie, A. (2014). *Carnivorous plants of Australia. Magnum opus.* Volumes 1 - 2, Doreset, England, Redferm Natural History.

Lowrie, A., Conran, J. G., & Moyle-Croft, J. (2002). A revision of *Byblis* (Byblidaceae) in south-western Australia. *Nuytsia, 15,* 11-19.

Lubertazzi, D., Lubertazzi, M. A. A., McCoy, N., Gove, A. D., Majer, J. D., & Dunn, R. R. (2010). The ecology of a keystone seed disperser, the ant *Rhytidoponera violacea. Journal of Insect Science,* 10:158.

Lughadha, E. N., & Proenca, C. (1996). A survey of the reproductive biology of the Myrtoideae (Myrtaceae). *Missouri Botantical Garden, 83,* 480-503.

Majer, J. D., & Lamont, B. B. (1985). Removal by ants of seed of *Grevillea pteridifolia* (Proteaceae). *Australian Journal of Botany, 33,* 611-618.

Majer, J. D., Gove, A. D., Sochacki, S., Searle, P., & Portloci, C. (2011). A comparison of the autecology of two seed-taking ant genera, *Rhytidoponera* and *Melophorus. Insects Sociaux, 58,* 115-25.

Mant, J. G., Schiesti, F. P., Peakall, R., & Weston, P. H. (2002). A phylogenetic study of pollinator conservatism among sexually deceptive orchids. *Evolution, 56,* 888-898.

Marginson, R., Sedgley, M., Douglas, T. J., & Knox, R. B. (1985). Structure and secretion of the extrafloral nectaries of Australian acacias. *Israel Journal of Botany, 34,* 91-102.

Maslin, B. R. (2014). Four new species of *Acacia* (Fabaceae: Mimosoideae) with fasciculate phyllodes from south-west Western Australia. *Nuytsia, 24,* 161-175.

Matthews, M. L., Gardner, J., & Sedgley, M. (1999). The proteaceous pistil: morphology and anatomy aspects of the pollen presenter and style of eight species across five genera. *Annals of Botany, 83,* 385-399.

McComb, J. A. (1966). The sex forms of species in the flora of the south-west of Western Australia. *Australian Journal of Botany, 14,* 303-316.

McLuckie, J. (1924). Studies in parasitism. I. A contribution to the physiology of the genus *Cassytha.* Part 1. *Proceedings of the Linnaean Society of New South Wales, 49,* 55-78.

Meney, K. A. & Pate, J. S. (1999) Australian Rushes. University of Western Australian Press, Nedlands.

Menz, M. H. M., Phillip, R. D., Dixon, K. W. Peakall, R., & Didham, R. K. (2013). Mate-searching behaviour of common and rare wasps and the implications for pollen movement of the sexually deceptive orchids they pollinate. *PLoS One, 8,* e59111.

Mercer, G. N., Gill, A. M., & Weber, R. O. (1994). A time-dependent model of fire impact on seed survival in woody fruits. *Australian Journal of Botany 42,* 71-81.

Merritt, D. J., Turner, S. R., Clarke, S., & Dixon, K. W. (2007). Seed dormancy and germination stimulation syndromes for Australian temperate species. *Australian Journal of Botany, 55,* 336-344.

Merwin, L., He, T., & Lamont, B. B. (2012). Phylogenetic and phenotypic structure among *Banksia* communities in south-western Australia. *Journal of Biogeography, 39,* 397-407.

Merwin, L, He, T., Lamont, B. B., Enright, N. J., & Krauss, S. L. (2012). Low rate of between-population seed dispersal restricts genetic connectivity and metapopulation dynamics in a clonal shrub. *PLoS ONE, 7,* e50974.

Midgley, J. J. (2000). What are the relative costs, limits and correlates of increased degree of serotiny? *Austral Ecology, 25,* 65-68.

Midgley, J. J. (2013). Flammability is not selected for, it emerges. *Australian Journal of Botany, 61,* 102-106.

Midgley, J. J., Cowling, R. M., & Lamont, B.B. (1991). Relationship of follicle size and seed size in *Hakea* (Proteaceae): isometry, allometry and adaptation. *South African Journal of Botany, 57,* 107-110.

Milberg, P. & Lamont, B. B. (1997). Seed/cotyledon size and nutrient content play a major role in early performance of species on nutrient-poor soils. *New Phytologist, 136,* 665-672.

Milberg, P., Pérez-Fernández, M. A., & Lamont, B. B. (1998). Seedling growth response to added nutrients depends on seed size in three woody genera. *Journal of Ecology, 86,* 624-632.

Miller, J. T., Andrew, R., & Bayer, R. J. (2003). Molecular phylogenetics of the Australian acacias of subg. *Phyllodineae* (Fabaceae: Mimosoideae) based on the *trn*K intron. *Australian Journal of Botany, 51,* 167-177.

Miller, J. T., & Miller, C. (2011). *Acacia* seedling morphology: phyllotaxy and its relationship to seed mass. *Australian Journal of Botany, 59,* 185-196.

Miller, J. T., Murphy, D. J., Ho, S. Y. W., Cantrill, D. J., & Seigler, D. (2013). Comparative dating of *Acacia*: combining fossils and multiple phylogenies to infer ages of clades with poor fossil records. *Australian Journal of Botany, 61,* 436-445.

Mitchell, P. J., Veneklass, E. J., Lambers, H., & Burgess, S. S. (2008a). Using multiple trait associations to define hydraulic functional types in lant communities of south-western Australia. *Oecologia, 158,* 385-397.

Mitchell, P.J., Veneklass, E. J., Lambers, H., & Burgess, S. S. (2008b). Leaf water relations during summer water deficits: differential responses in turgor maintenance and variation in leaf structure among different plant communities in south-western Australia. *Plant, Cell and Environment, 31,* 1791-1802.

Moran, G. F., & Hopper, S. D. (1983). Genetic diversity and the insular population structure of the rare granite rock species, *Eucalyptus caesia* Benth. *Australian Journal of Botany, 31,* 161-172.

Morcombe, M. K. (1968). *Australia's western wildflowers*. Perth: Western Australia: Morcombe.

Mucina, L., & Majer, J. D. (2012). Ants and the origins of plant diversity in old, climatically stable landscapes: a great role for tiny players. *South African Journal of Botany, 83,* 44-46.

Mucina, L., Laliberté, E., Thiele, K. R., Dodson, J. R., & Harvey, J. (2014). Biogeography of kwongan: origins, diversity, endemism and vegetation patterns. In H. Lambers (Ed), *Plant life on the sandplains in southwest Australia.* pp. 38-80. Perth, Australia: UWA Publishing.

Mucina, L., & Wardell-Johnson, G. W. (2011). Landscape age and soil fertility, climatic stability, and fire regime predictability: beyond the OCBIL framework. *Plant and Soil, 341,* 1-23.

Murphy, D. J., Brown, G. K., Miller, J. T., & Ladiges, P. Y. (2010). Molecular phylogeny of *Acacia* Mill. (Mimosoideae: Leguminosae): evidence for major clades and informal classification. *Taxon, 59,* 7-19.

Mutch, R. W. (1970). Wildland fires and ecosystems – a hypothesis. *Ecology, 51,* 1046-1051.

Myers, N., Mittermeier, R. A., Mitterneier, C. G., da Fonseca, G. A. B., & Kent, J. (2000). Biodiversity hotspots for conservation priorities. *Nature,* 403, 853-858.

Navarro, E., Rouvier, C., Normand, P., Domenach, A. M., Simonet, P., & Prin, Y. (1998). Evolution of *Frankia*-Casuarinaceae interactions. *Genetics Selection Evolution, 30,* S357-S372.

Nicolle, D. (2006). A classification and census of regenerative strategies in the eucalypts (*Angophora, Corymbia* and *Eucalyptus* – Myrtaceae), with special reference to the obligate seeders. *Australian Journal of Botany, 54,* 391-407.

Nicotra, A. B., Cosgrove, M. J., Cowling, A., Schlichting, C. D., & Jones, C. S. (2008). Leaf shape linked to photosyntethic rates and temperature optima in South African *Pelargonium* species. *Oecologia, 154,* 625-635.

Niinemets, Ü. (2001). Global-scale climatic controls of leaf dry mass per area, density, and thickness in trees and shrubs. *Ecology, 82,* 453-469.

Noss, R. F. (1990). Indicators for monitoring biodiversity: s hierarchical approach. *Conservation Biology, 4,* 355-364.

Nyári, A., Peterson, A. T., Rice, N. H. & Moyle, R. G. (2009). Phylogenetic relationships of flower-peckers (Aves: Dicaeidae): novel insights into the evolution of a tropical passerine clade. *Molecular Phylogenetics and Evolution, 53,* 613-619.

Obrien, S. P. (1995). Extrafloral nectaries in *Chamelaucium uncinatum*: a first record in the Myrtaceae. *Australian Journal of Botany*, *43*, 407-13.

Parkes, D. M., & Hallam, N. D. (1984). Adaptation for carnivoryin the West Australian pitcher plant *Cephalotus follicularis* Labill. *Australian Journal of Botany, 32*, 595-604.

Parsons, M. H., Koch, J. M., Lamont, B. B., Vlahos, S., & Fairbanks, M. (2006). Planting density effects and selective herbivory by kangaroos on species used in restoring forest communities. *Forest Ecology and Management, 229*, 39-49.

Parsons, M. H., Lamont, B. B., Davies S. J. J. F., & Kovacs B. R., (2005). How energy and co-available foods affect the currency of forage for the western grey kangaroo. *Animal Behaviour, 71*, 765-772.

Parsons, R. F., & Hopper, S. D. (2003). Monocotyledonous geophytes: comparison of south-western Australia with other areas of mediterranean climate. *Australian Journal of Botany, 51*, 129-133.

Pate, J. S. (1989). Australian micro stilt plants. *Trends in Ecology and Evolution, 4*, 45-49.

Pate, J. S. (2002). Symbiotic nitrogen fixation between microorganisms and higher plants of natural ecosystems. In K. Sivasithamparam, K. W. Dixon & R. L. Barrett (Eds), *Microogranisms in plant conservation and biodiversity*, pp. 44-77. The Hague: Kluwer Academic Publishers.

Pate, J. S., Davidson, N. J., Kuo, J., & Milburn, J. A. (1990a). Water relations of the root hemiparasite *Olax phyllanthi* (Labill.) R.Br. (Olacaceae) and its multiple hosts. *Oecologia, 84*, 186-193.

Pate, J. S., & Dixon, K. W. (1982). *Tuberous, cormous and bulbous plants – biology of an adaptive strategy in Western Australia.* Nedlands, Australia: University of Western Australia Press.

Pate, J. S., & Dixon, K. W. (1996). Convergence and divergence in the southwestern Australian flora in adaptations of roots to limited availability of water and nutrients, fire and heat stress. In S. D. Hopper, M. Harvey, J. Chappill and A. S. George (Eds), *Gondwanan heritage: Past, present and future of the Western Australian biota*, pp. 249-258. Chipping Norton: Surrey Beatty and Sons.

Pate, J. S., Dixon, K. W. & Orshan, G. (1984). Growth and life form characteristics of kwongan species. In J. S. Pate and J. S. Beard (Eds), *Kwongan. Plant life of the sandplain* pp. 84-100. Nedlands, Australia: University of Western Australia Press.

Pate, J. S., Kuo, J., & Davidson, N. J. (1990b). Morphology and anatomy of the haustorium of the root hemiparasite *Olax phyllanthi* (Olacaceae) with special reference to the haustorial interface. *Annals of Botany*, 65, 425-436.

Pate, J. S., Kuo, J., Dixon, K. W., & Crisp, M. D. (1989). Anomalous secondary thickening in roots of *Daviesia* (Fabaceae) and its taxonomic significance. *Botanical Journal of the Linnaean Society, 99*, 175-193.

Pate, J. S., Pate, S. R., Kuo, J., & Davidson, N. J. (1990b). Growth, resource allocation and haustorial biology of the root hemiparasite *Olax phyllanthi* (Olacaceae). *Annals of Botany, 65*, 437-449.

Pate, J. S., Rasins, E., Rullo, J., & Kuo, J. (1986). Seed nutrient reserves of Proteaceae with special reference to protein bodies and their inclusions. *Annals of Botany, 57*, 747-770.

Pate, J. S., Rasins, E. , Thumfort, P. P., McCheseney, C. J. & Meney, K. A. (1995). Occurrence of the unusual amino acid compound S-methylcysteine in Australian members of the Restionaceae: biological and taxonomic significance. *Australian Journal of Botany,* 43, 73-84.

Pate, J. S., Veerboom, W. H., & Galloway, P. D. (2001). Co-occurrence of Proteaceae, laterite and related oligotrophic soils: coincidental associations or causative inter-relationships? *Australian Journal of Botany, 49*, 529-560.

Pate, J. S., Weber, G., & Dixon, K. W. (1984). Stilt plants – Extraordinary growth form of the kwongan. In J. S. Pate and J. S. Beard (Eds), *Kwongan. Plant life of the sandplain* pp. 101-125. Nedlands, Australia: University of Western Australia Press.

Pausas, J. G., Bradstock, R. A., Keith, D. A., Keeley, J. E., & GCTE. (2004). Plant functional traits in relation to fire in crown-fire ecosystems. *Ecology, 85*, 1085-1100.

Pausas, J. G., & Keeley, J. E. (2009). A burning story: the role of fire in the history of life. *BioScience, 59*, 593–601.

Pausas, J. G., & Keeley, J. E. (2014). Evolutionary ecology of resprouting and seeding in fire-prone ecosystems. *New Phytologist, 204*, 55-65.

Peakall, R. (1989). The unique pollination of *Leporella fimbriata* (Orchidaceae) by pseudocopulationg winged male ants *Myrmecia urens* (Formicidae). *Plant Systematics and Evolution, 167,* 137-148.

Peakall, R. (1990). Responses of male *Zaspilothynnus trilobatus* Turner wasps to females and the sexually deceptive orchid it pollinates. *Functional Ecology, 4,* 159–167.

Peakall, R., Angus, C. J., & Beattie, A. J. (1990). The significance of ant and plant traits for ant pollination in *Leporella fimbriata. Oecologia, 84,* 457-460.

Peakall, R., & Beattie, A. J. (1991). The genetic consequences of worker ant pollination in a self-compatible, clonal orchid. *Evolution, 45,* 1837-1848.

Peakall, R., & Beattie, A. J. (1996). Ecological and genetic consequences of pollination by sexual deception in the orchid *Caladenia tentaculata. Evolution, 50,* 2207-2220.

Peakall, R., Ebert, D., Poldy, J., Barrow, R. A., Francke, W., Bower, C. C., & Schiestl, F. P. (2010). Pollinator specificity, floral odour chemistry and the phylogeny of Australian sexually deceptive *Chiloglottis* orchids: implications for pollinator-driven speciation. *New Phytologist, 188,* 437-450.

Peakall, R., & James, S. H. (1989). Outcrossing in an ant pollinated clonal orchid. *Heredity 62,* 161-167.

Pérez-Harguindeguy, N. M., Díaz, S., Garnier, E., Lavorel, S., Poorter, H., *et al.*, (2013). New handbook for standardised measurement of plant functional traits worldwide. *Australian Journal of Botany, 61,* 167-234.

Pestell, A. J. L, & Petit, S. (2007). Diet of the western pygmy possum, *Cercartetus concinnus* Gould (Marsupialia: Burramyidae), at Innes National Park, South Australia, and evaluation of diet sampling methods. *Australian Journal of Zoology, 55,* 275-284.

Phillips, R. D., Barrett, M. D., Dixon, K. W., & Hopper, S. D. (2011). Do mycorrhizal symbioses cause rarity in orchids? *Journal of Ecology, 99,* 858-869.

Philips, R. D., Brown, A. P., Dixon, K. W., & Hopper, S. D. (2011). Orchid biogeography and factors associated with rarity in a biodiversity hotspot, the Southwest Australian Florisitic Region. *Journal of Biogeography, 38,* 487-501.

Phillips, R. D., Faast, R., Bower, C. C., Brown, G. R., & Peakall, R. (2009). Implications of pollination by food and sexual deception for pollinator specificity, fruit set, population genetics and conservation of *Caladenia* (Orchidaceae). *Australian Journal of Botany, 57,* 287-306.

Phillips, R. D., Hopper, S. D., & Dixon, K. W. (2010). Pollination ecology and the possible impacts of environmental change in the southwest Australian biodiversity hotspot. *Philosophical Transactions of the Royal Society B, 365,* 517-528.

Phillips, R. D., Scaccabarozzi, D., Retter, B. A., Hayes, C., Brown, G. R., Dixon, K. W., & Peakall, R. (2014). Caught in the act: pollination of sexually deceptive trap-flowers by fungus gnats in *Pterostylis* (Orchidaceae). *Annals of Botany, 113,* 629-641.

Pierce, N. B., & Simpson, M. G. (2009). Polyapertutate pollen types and ratios of heteromorphism in the monocot genus *Conostylis* (Haemodoraceae). *Australian Systematic Botany, 22,* 16-30.

Playsted, C. W. S., Johnston, M. E., Ramage, C. M., Edwards, D. G., Cawthray, G. R., & Lambers, H. (2006). Functional significance of dauciform roots: exudation of carboxylates and acid phosphatase under phosphorus deficiency in *Caustis blakei* (Cyperaceae). *New Phytologist, 170,* 491-500.

Poot, P., Hopper, S. D., & van Diggelen, J. M. H. (2012). Exploring rock fissures: does a specialized root morphology explain endemism on granite outcrops? *Annals of Botany, 110,* 291-300.

Poot, P., & Lambers, H. (2008). Shallow-soil endemics: adaptive advantages and constraints of a specialized root-system morphology. *New Phytologist 178,* 371-381.

Poot, P., & Veneklaas, E. J. (2013). Species distribution and crown decline are associated with contrasting water relations in four common sympatric eucalypt species in southwestern Australia. *Plant and Soil, 364,* 409-423.

Poorter, H, Niinemets, Ü., Poorter, L., Wright, I.J., & Villar, R. (2009). Causes and consequences of variation in leaf mass per area (LMA): a meta-analysis. *New Phytologist, 182,* 565-588.

Prideaux, G. J., & Warburton, N. M. (2010). An osteology-based appraisal of the phylogeny and evolution of kangaroos and wallabies (Macropodidae: Marsupialia). *Zoological Journal of the Linnean Society, 159,* 954-987.

Rafferty, C., & Lamont, B. B. (2007). Selective herbivory by mammals on 19 species planted at two densities. *Acta Oecologica, 32,* 1-13.

Rafferty, C. M., Lamont, B. B., & Hanley, M. E. (2005). Selective feeding by kangaroos (*Macropus fuliginosus*) on seedlings of *Hakea* species: effects of chemical and physical defences. *Plant Ecology, 177,* 201-208.

Rafferty, C. M., Lamont, B. B., & Hanley, M. E. (2010). Herbivore feeding preferences in captive and wild populations. *Austral Ecology, 35,* 257-263.

Rasmussen, H. N., & Rasmussen, F. N. (2009). Orchid mycorrhiza: implications of a mycophagous life style. *Oikos, 118,* 334-345.

Read, J., Sanson, G. D., & Lamont, B. B. (2005). Leaf mechanical properties in sclerophyll woodland and shrubland on contrasting soils. *Plant and Soil, 276,* 95-113.

Reich, P. B. (2014). The world-wide 'fast-slow' plant economics spectrum: a traits manifesto. *Journal of Ecology, 102,* 275-301.

Reid, N. (1987). The mistletoebird and Australian Mistletoes: co-evolution or coincidence? *Emu, 87,* 130-131.

Richards, M. B., & Lamont, B. B. (1996). Post-fire mortality and water relations of three congeneric shrub species under extreme water stress – a trade-off with fecundity? *Oecologia, 107,* 53-60.

Richardson, K. C., Wooller, R. D., & Collins, B. G. (1986). Adaptations to a diet of nectar and pollen in the marsupial *Tarsipes rostratus* (Marsupialia: Tarsipedidae). *Journal of Zoology, 208,* 285-297.

Rivadavia, F., Kondo, K., Kato, M. & Hasebe, M. (2003). Phylogeny of the sundews, *Drosera* (Droseraceae), based on chloroplast *rbcL* and nuclear 18S ribosomal DNA sequences. *American Journal of Botany, 90,* 120-130.

Roche, S., Dixon, K. W., & Pate, J. S. (1997). Seed ageing and smoke: partner cues in the amelioration of seed dormancy in selected Australian native species. *Australian Journal of Botany, 45,* 783-815.

Roche, S., Dixon, K. W., & Pate, J. S. (1998). For everything a season: smoke-induced seed germination and seedling recruitment in a Western Australian *Banksia* woodland. *Australian Journal of Ecology, 23,* 111-120.

Rognon, P., & Williams, M. A. J. (1977). Late Quaternary climatic changes in Australia and North Africa: a preliminary interpretation. *Palaeogeography, Palaeoclimatology, Palaeocology, 21,* 285-327.

Rokich, D. P., Meney, K. A., Dixon, K. W., &. Sivasithamparam, K. (2001). The impact of soil disturbance on root development in woodland communities in Western Australia, *Australian Journal of Botany, 49,* 169-183.

Rossetto, M., Jezierski, G., Hopper, S. D., & Dixon, K. W. (1999). Conservation genetics and clonality in two critically endangered eucalypts from the highly endemic south-western Australian flora. *Biological Conservation, 88,* 321-331.

Roth-Nebelsick, A., Hassiotou, F., & Veneklaas, E. J. (2009). Stomatal crypts have small effects on transpiration: A numerical model analysis *Plant Physiology, 151,* 2018–2027.

Rourke, J., & Wiens, D. (1977). Convergent floral evolution in South African and Australian Proteaceae and its possible bearing on pollination by nonflying mammals. *Annals of the Missouri Botanical Garden, 64,* 1-17.

Saffer, V. M. (2004). Are diel patterns of nectar production and anthesis associated with other floral traits in plants visited by potential bird and mammal pollinators? *Australian Journal of Botany, 52,* 87-92.

Sampson, J. F., Hopper, S. D., & James, S. H. (1989). The mating system and population genetic structure in a bird-pollinated mallee, *Eucalyptus rhodantha. Heredity, 63,* 383-393.

Sauquet, H., Weston, P. H., Anderson, C. L., Barker, N. P., Cantrill, D. J., Mast, A. R., & Savolainen, V. (2009). Contrasted patterns of hyperdiversification in Mediterranean hotspots. *Proceedings of the National Academy of Sciences, 106,* 221–225.

Sedgley, M., Sierp, M., Wallwork, M. A., Fuss, A. M., & Thiele, K. (1993). Pollen presenter and pollen morphology of *Banksia* L.f. (Proteaceae). *Australian Journal of Botany, 41,* 439-464.

Schiestl, F. P. (2005). On the success of a swindle: pollination by deception in orchids. *Naturwissenschaften, 92,* 255–264.

Schiestl, F. P., Peakall, R., & Mant, J. (2004). Chemical communication in the sexually deceptive orchid genus *Cryptostylis. Botanical Journal of the Linnaean Society, 144,* 199–205.

Schiestl, F. P., & Schlüter, P. M. (2009). Floral isolation, specialized pollination, and pollinator behavior in orchids. *Annual Review of Entomology, 54,* 425-446.

Schulze, E. –D., Gebauer, G., Schulze, W., & Pate, J. S. (1991). The utilization of nitrogen from insect capture by different growth forms of *Drosera* from southwest Australia. *Oecologia, 87,* 240-246.

Schulze, W., Schulze, E.-D., Pate, J.S., & Gillison, A. N. (1997). The nitrogen supply from soils and insects during growth of the pitcher plant *Nepenthes mirabilis, Cephalotus follicularis* and *Darlingtonia californica. Oecologia, 112,* 464-471.

Schut, A. G. T., Wardell-Johnson, G. W., Yates, C. J., Keppel, G., Baran, I., Franklin, S. E., Hooper, S. D., van Niel. K. P., Mucina, L., & Byrne, M. (2014). Rapid characterisation of vegetation structure to predict refugia and climate change impacts across a global biodiversity hotspot. P*LoS ONE, 9,* e82778.

Scopece, G., Cozzolino, S., Johnson, S. D., & Schiestl, F. P. (2010). Pollination efficiency and the evolution of apecialized seceptive pollination systems. *The American Naturalist*, 175, 98-105.

Shane, M. W., Cawthray, G. R., Cramer, M. D., Kuo, J., & Lambers, H. (2006). Specialised 'dauciform' roots of Cyperaceae are structurally distinct, but functionally analogous with 'cluster' roots. *Plant, Cell and Environment,* 29, 1989-1999.

Shane, M. W., Dixon, K. W., & Lambers, H. (2005). The occurrence of dauciform roots amongst Western Australian reeds, rushes and sedges, and the impact of phosphorus supply on dauciform-root development in *Schoenus unispiculatus* (Cyperaceae). *New Phytologist, 165,* 887-898.

Shane, M. W., McCully, M. E., Canny, M. J., Pate. J. S., Ngo, H., Mathesius, U., Cawthray, G. R., & Lambers, H. (2009). Summer dormancy and witner growth: root survival in a perennial monocotyledon. *New Phytologist, 183,* 1085-1096.

Shane, M. W., McCully, M. E., Canny, M. J., Pate. J. S., & Lambers, H. (2011). Development and persistence of sandsheaths of *Lyginia barbata* (Restionaceae): relation to root structural development and longevity. *Annals of Botany, 108,* 1307-1322.

Shen, H., Prider, J. N., Facelli, J. M., & Watling, J. R. (2010). The influence of the hemiparasitic angiosperm *Cassytha pubescens* on photosynthesis of its host *Cytisus scoparius. Functional Plant Biology, 37,* 14-21.

Shen, H., Ye, W., Hong, L., Cao, H., & Wang, Z. (2005). Influence of the obligate parasite *Cuscuta campestris* on growth and biomass allocation of its host *Mikania micrantha. Journal of Experimental Botany, 56,* 1277-1284.

Shrestha, M., Dyer, A. G., Body-Gemy, S, Wong, B. B. M., & Burd, M. (2013). Shades of red: bird-pollinated flowers target the specific colour discrimination abilities of avian vision. *New Phytologist, 198,* 301-310.

Smith, M. A., Bell, D. T., & Loneragan, W. A. (1999). Comparative seed germination ecology of *Austrostipa compressa* and *Ehrharta calycina* (Poaceae) in a Western Australian *Banksia* woodland. *Australian Journal of Ecology, 24,* 35-42.

Smith, R. J., Hopper, S. D., & Shane, M. W. (2011). Sand-binding roots in Haemodoraceae: global survey and morphology in a phylogenetic context. *Plant and Soil, 348,* 453-470.

Smith, W. K., Bell, D. T., & Shepherd, K. A. (1998). Associations between leaf structure, orientation, and sunlight exposure in five Western Australian communities. *American Journal of Botany, 85,* 56-63.

Sommer, B. & Froend, R. (2011). Resilence of phreatophytic vegetation to groundwater drawdown: is recovery possible under a drying climate? *Ecohydrology, 4,* 67-82.

Sommer, B., & Froend, R. (2014). Phreatophytic vegetation responses to groundwater depth in a drying mediterranean-type landscape. *Journal of Vegetation Science, 25,* 1045-1055.

Sprent, J. I., (2007). Evolving ideas of legume evolution and diversity: a taxonomic perspective on the occurrence of nodulation. *New Phytologist, 174,* 11-25.

Stace, H. M. (1995). Protogyny, self-incompatibility and pollination in *Anthocercis gracilis* (Solanaceae) *Australian Journal of Botany, 43,* 451-459.

Stoutamire, W. P. (1983). Wasp-pollinated species of *Caladenia* (Orchidaceae) in south-western Australia. *Australian Journal of Botany, 31,* 383-394.

Sydes, M. A., & Calder, D. M. (1993). Comparative reproductive biology of two sun-orchids, the vulnerable *Thelymitra circumsepta* and the widespread *T. ixioides* (Orchidaceae). *Australian Journal of Botany, 41,* 577-589.

Swarts, N. D., & Dixon, K. W. (2009). Terrestrial orchid conservsation in the age of extinction. *New Phytologist, 104,* 543-556.

Swarts, N. D., Sinclair, E. A., Francis, A. & Dixon, K. W. (2009). Ecological speciation in mycorrhizal symbiosis leads to rarity in an endangered orchid. *Molecular Ecology, 19,* 3226-3242.

Tapper, S. –L., Byrne, M., Yates, C. J., Keppel, G., Hopper, S. D., van Niel, K., Schut, A. G. T., Mucina, L, & Wardell-Johnson, G. W. (2014). Isolated with persistence or dynamically connected? Genetic patterns in a common granite outcrop endemic. *Diversity and Distributions, 20,* 987-1001.

Tennakoon, K. U., Pate, J. S., & Arthur, D. (1997b). Ecophysiological aspects of the woody root hemiparasite *Santalum acuminatum* (R.Br.) A.DC and its common hosts in south western Australia. *Annals of Botany, 80,* 245-256.

Tennakoon, K. U., Pate, J. S., & Fineran B. A. (1997a). Growth and partitioning of C and N in the shrub legume *Acacia littorea* in presence or absence of the root hemiparasite *Olax phyllanthi*. *Journal of Experimental Botany, 48,* 1061-1069.

The Angiosperm Phylogeny Group (2009). An update of the Angiosperm Phylogeny Group classification for the orders and families of flowering plants: APG III. *Botanical Journal of the Linnean Society, 161,* 105-121.

Thiele, K. R., Wylie, S. J., Maccarone, L., & Hollick, P. (2008). *Pilostyles coccoidea* (Apodanthaceae), a new species from Western Australia described from morphological and molecular evidence. *Nuytsia, 18,* 273-284.

Thornhill, A. H., Wilson, P. G., Drudge, J., Barrett, M. D., Hope, G. S., Craven, L. A., & Crisp, M. D. (2012). Pollen morphology of the Myrtaceae. Part 3. Tribes Chamelaucieae, Leptospermeae and Lindsayomyrteae. *Australian Journal of Botany, 60,* 225-259.

Tieu, A., Dixon, K. W., Meney, K. A., & Sivasithamparam, K. (2001a). The interaction of heat and smoke in the release of seed dormancy in seven species from southwestern Western Australia. *Annals of Botany, 88,* 259-265.

Tieu, A., Dixon, K. W., Meney, K. A., & Sivasithamparam, K. (2001b). Interaction of soil burial and smoke on germination patterns is seeds of selected Australian native plants. *Seed Science Research, 11,* 69-76.

Tieu, A., & Egerton-Warburton, L. M. (2000). Contrasting seed morphology dynamics in relation to the alleviation of dormancy with soil storage. *Canadian Journal of Botany, 78,* 1187-1198.

Toon, A., Cook, L. G., & Crisp, M. D. (2014). Evolutionary consequences of shifts to bird-pollination in the Australian pea-flowered legumes (Mirbelieae and Bossiaeeae). *BMC Evolutionary Biology, 14,* 43.

Turner, K. M., & Frederickson, M. E. (2013). Signals can trump rewards in attracting seed-dispersing ants. *PLoS ONE, 8,* e71871.

Twigg, L. E., King, D. R., Bowen, L. H., Wright, G. R., & Eason, C. T. (1996). Fluoroacetate content of some species of the toxic Australian plant genus, *Gastrolobium*, and its environmental persistence. *Natural Toxins, 4,* 122-127.

Twigg, L. E., & Socha, L. V. (1996). Physical versus chemical defence mechanisms in toxic *Gastrolobium*. *Oecologia, 108,* 21-28.

Tucker, S. C., & Bernhardt, P. (2000). Floral ontogeny, pattern formation, and evolution in *Hibbertia* and *Adrastaea* (Dilleniaceae). *American Journal of Botany*, 87, *1915-1936.*

van Staden, J., Brown, N. A. C., Jäger, A. K., & Johnson, T. A. (2000). Smoke as a germination cue. *Plant Species Biology, 15,* 167-178.

Vanstone, V. A., & Paton, D. C. (1988). Extrafloral nectaries and pollination of *Acacia pycnantha* Benth by birds. *Australian Journal of Botany, 36*, 519-531.

Vaughton, G., & Ramsey, M. (2001). Variation in summer dormancy in the lilioid geophyte *Burchardia umbellata* (Colchicaceae). *American Journal of Botany, 88,* 1223-1229.

Veneklass, E. J., & Poot, P. (2003). Seasonal patterns in water use and leaf turnover of different plant functional types in a species-rich woodland, south-western Australia. *Plant and Soil 257,* 295-304.

Venterink, H. O. (2011). Does phosphorus limitation promote species-rich plant communities? *Plant and Soil, 345,* 1-9.

Vidal-Russell, R., & Nickrent, D. L. (2008a). The first mistletoes: origins of aerial parasitism in Santalales. *Molecular Phylogenetics and Evolution, 47,* 523-537.

Vidal-Russell, R., & Nickrent, D. L. (2008b). Evolutionary relationships in the showy mistletoe family (Loranthaceae). *American Journal of Botany, 95,* 1015-1029.

Walker, B. A., & Pate, J. S., (1986). Morphological varation between seedling progenies of *Viminaria juncea* (Schrad. & Wendl.) Hoffmans. (Fabaceae) and its physiological significance. *Australian Journal of Plant Physiology, 13,* 305-319.

Ward, S. C., Koch, J. M., & Grant, C. D. (1997). Ecological aspects of soil seed-banks in relation to bauxite mining. I. Unmined jarrah forest. *Australian Journal of Ecology, 22,* 169-176.

Wardell-Johnson, G., & Coates, D. (1996). Links to the past: local endemism in four species of forest eucalypts in southwest Australia. In S. D. Hopper, M. Harvey, J. Chappill and A. S. George (Eds), *Gondwanan heritage: Past, present and future of the Western Australian biota*, pp.137-154. Chipping Norton: Surrey Beatty and Sons.

Wedge, J. A. (2012). Navigating the floral Milky Way: the taxonomy of the microgeophytic triggerplants (*Stylidium petiolare* and allies, Stylidiaceae). *Australian Systematic Botany, 25,* 138-169.

Weiss, R.W., & Lamont B.B. (1997). Floral color change and insect pollination: a dynamic relationship. *Israel Journal of Plant Sciences, 45,* 185-199

Westcott, V. C., Enright, N. J., Miller, B. P., Fontaine, J. B., Lade, J. C., & Lamont, B. B. (2014). Biomass and litter accumulation patterns in species-rich shrublands for fire hazard assessment. *International Journal of Wildland Fire, 23,* 860-871.

Whelan, R. J., & Burbidge, A. H. (1980). Flowering phenology, seed set and bird pollination of five Western Australian *Banksia* species *Australian Journal of Ecology, 5,* 1–7.

White, N. E., Phillips, M. J., Gilbert, M. T., Alfaro-Nũñez, A., Willerslev, E., Mawson, P. R., Specier, P.B. S., & Bunce, M. (2011). The evolutionary history of cockatoos (Aves: Psittaciformes: Cacatuidae). *Molecular Phylogenetics and Evolution*, 59, 615-622.

White, R. H., & Zipper, W. C. (2010). Testing and classification of individual plants for fire behaviour: plant selection for the wildland-urban interface. *International Journal of Wildland Fire, 19,* 213-227.

Wiens, D., Renfree, M., & Wooller, R. O. (1979). Pollen loads of honey possums (*Tarsipes spenserae*) and nonflying mammal pollination in southwestern Australia. *Annals of the Missouri Botanical Garden, 66,* 830-838.

Wilson, C. A., & Calvin, C. L. (2006). An origin of aerial branch parasitism in the mistletoe family, Loranthaceae. *American Journal of Botany, 93,* 787-796.

Witkowski, E.T. F., & Lamont, B. B. (1991). Leaf specific mass confounds leaf density and thickness. *Oecologia 88,* 486-493.

Witkowski, E. T. F., & Lamont, B. B. (1996). Disproportionate allocation of mineral nutrients and carbon between vegetative and reproductive structures in *Banksia hookeriana. Oecologia, 105,* 38-42.

Witkowski, E. T. F., & Lamont, B. B. (1997). Does the rare *Banksia goodii* have inferior vegetative, reproductive or ecological attributes compared with its widespread co-occurring relative *B. gardneri? Journal of Biogeography, 24,* 469-482.

Witkowski, E. T. F., Lamont, B. B., Walton, C. S., & Radford, S. (1992). Leaf demography, sclerophylly and ecophysiology of two *Banksia* species with contrasting leaf life spans. *Australian Journal of Botany, 40,* 849-862.

Wooller, R. D., Renfree, M. B., Russell, E. M., Dunning, A., Green, S. W., & Duncan, P. (1981). Seasonal changes in a population of the nectar-feeding marsupial *Tarsipes spencerae* (Marsupialia: Tarsipedidae). *Journal of Zoology, 195,* 267-279.

Wooller, R. D., Russell, E. M. Renfree, M. B., & Towers, P. A. (1984). A comparison of seasonal changes in the pollen poads of nectarivorous marsupials [*Tarsipes*] and birds [honeyeaters]. *Australian Wildlife Research* 10, 311 – 317.

Wooller, R. D., & Wooller, S. J. (2003). The role of non-flying animals in the pollination of *Banksia nutans. Australian Journal of Botany, 51,* 503-507

Wright, I. J., Reich, P. B., Westoby, M. *et al.* (2004). The worldwide leaf economics spectrum. *Nature, 428,* 821-827.

Yan, Z.G. (1990). Host specificity of *Lysiana exocarpi* subsp *exocarpi* and other mistletoes in southern South Australia. *Australian Journal of Botany, 38,* 475-486.

Yan, Z. (1993a). Seed dispersal of *Amyema preissii* and *Lysiana exocarpi* by mistletoe birds and spiny-cheeked honeyeaters. *Emu, 93,* 214-219.

Yan, Z. (1993b). Germination and seedling development of two mistletoes, *Amyema preissii* and *Lysiana exocarpi*: host specificity and mistletoe-host compatability. *Australian Journal of Ecology, 18,* 419-429.

Yan, Z. (1993c). Resistance to haustorial development of two mistletoes, *Amyema preissii* (Miq.) Tieghem and *Lysiana exocarpi* (Behr.) Tieghem ssp. *Exocarpi* (Loranthaceae), on host and nonhost species. *International Journal of Plant Sciences, 154,* 386-394.

Yates, C. J., Elliott, C., Byrne, M, Coates, D. J., & Fairman R. (2007). Seed production, germinability and seedling growth for a bird pollinated shrub in fragments of kwongan in south-west Australia. *Biological Conservation, 136,* 306-314.

Yesson, C., & Culham, A. (2006). Phyloclimatic modelling: combining phylogenetics and bioclimatic modelling. *Systematic Biology, 55,* 785-802.

Zencich, S. J., Froend, R.H., Turner, J. V., & Galilitis, V. (2002). Influence of groundwater depth on the seasonal sources of water accessed by *Banksia* tree species on a shallow, sandy coastal aquifer. *Oecologia, 131,* 8-19.

About the Authors

Dr Philip Groom

Philip was a Lecturer in Environmental Biology at Curtin University, Western Australia for seven years, specialising in teaching and researching on the ecology of the southwestern Australian flora, especially strategies to survive the fire-prone, hot, dry summers. He has conducted research on post-fire recovery, drought responses, groundwater dependency, leaf morphology and biogeography of the flora, focusing mostly on the sandplain vegetation. He has a great passion for the ecology of the genus *Hakea* (Proteaceae) that served as the focus of his PhD, awarded in 1996.

Emeritus Professor Byron Lamont

Byron is a Distinguished Professor (Emeritus) in Plant Ecology at Curtin University. He has been studying the ecology and ecophysiology of the southwestern Australian biota since 1969, pioneering experimental work on specialised (proteoid) roots and canopy seed storage (serotiny) in the Proteaceae. Byron has undertaken research on the nutrient/water relations, reproductive biology, population dynamics, species diversity, evolution and management of the southwestern Australian flora. He was awarded a Doctor of Science for this work by the University of Western Australia in 1993. Prof Lamont is regarded as the father of ecological studies on the southwestern Australian sandplains. He was inaugural Director of the Centre for Ecosystem Diversity and Dynamics at Curtin. Byron was made a Member of the Order of Australia in 2010 for his services to the Australian flora as an educator, researcher and author. In 2012, he received the Clarke Medal from the Royal Society of NSW for distinguished work in the Natural Sciences.

List of Figures

List of Tables

Index

Bei Fragen zur Produktsicherheit wenden Sie sich bitte an:
If you have any questions regarding product safety,
please contact:

Walter de Gruyter GmbH
Genthiner Straße 13
10785 Berlin
productsafety@degruyterbrill.com